Nathaniel Lansing

7R

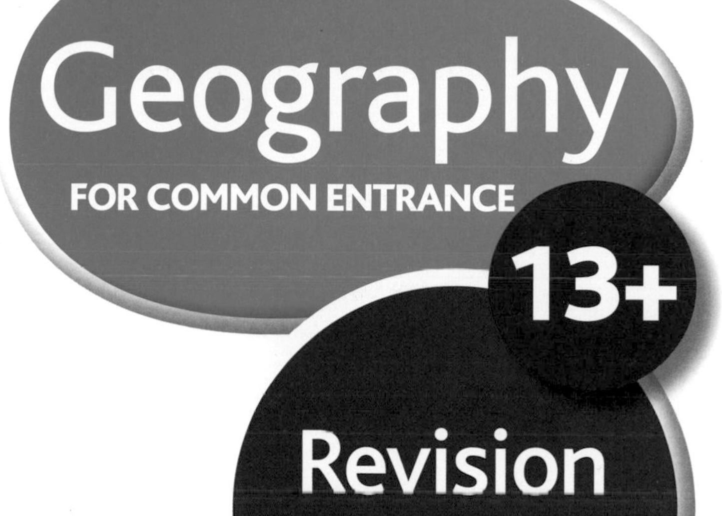

Geography

FOR COMMON ENTRANCE

13+

Revision Guide

Belinda Froud-Yannic
Subject Editor: Simon Lewis

About the author

Belinda Froud-Yannic has been Head of Geography at Thomas's School, Clapham since 2001. She formerly taught at Broomfield School, Southgate. Most of her spare time is taken up with looking after her two children but she also tries to squeeze in some pottery, walking up hills and skiing. She believes that Geography is a subject that everyone can enjoy due to its diversity of themes and that lessons should be fun and full of fieldwork.

Acknowledgements

I would like to dedicate this book to the pupils of Thomas's Clapham from 2001 to 2014. They have worked incredibly hard, achieving fantastic Common Entrance results and a huge percentage have reached the dizzy heights of Top Geographer status! Thanks go to them for their constant ideas on how the guide could be improved and updated. Also, sincere thanks go to Simon Lewis, my editor.

I would also like to thank Prue and Ernie for putting up with a very distracted mother while editing this book.

Orders: please contact Bookpoint Ltd, 130 Milton Park, Abingdon, Oxon OX14 4SB. Telephone: +44 (0)1235 827827. Lines are open 9.00a.m.–5.00p.m., Monday to Saturday, with a 24-hour message answering service. Visit our website at www.galorepark.co.uk for details of other revision guides for Common Entrance, examination papers and Galore Park publications.

Published by Galore Park Publishing Ltd
An Hachette UK company
Carmelite House, 50 Victoria Embankment, London EC4Y 0DZ
www.galorepark.co.uk

Impression number 10 9 8 7 6 5 4 3
2018 2017 2016

Typeset in India
Printed in Spain
Illustrations by Aptara, Inc.
A catalogue record for this title is available from the British Library.

ISBN: 978 1 471827 30 3

Contents

Introduction

I hope that you find this guide as useful as the pupils of Thomas's Clapham have over the past ten years. It will be a vital aid before your Common Entrance examination and will also be of use during lessons and for homework. Keep it with you as often as possible so that you can make the most of any free time you have.

Throughout the book, you will find the following:

- Exam-style questions – these are example Common Entrance questions for you to try during your revision period (the answers are near the back of the book).
- Make sure you know – these appear towards the end of each chapter and summarise what you should know on that subject.
- Test yourself questions – these appear at the end of each chapter and test what you know about that subject (the answers are near the back of the book). They also include a list of 'words you need to know' for the exam.
- Revision tips – these will give you ideas about how to revise particular subjects.

You need to know how to draw certain diagrams for the exam. As you work through this book, look out for notes in the margin which will tell you if you need to memorise a diagram.

If you are unsure of any topic area remember to ask your teacher for help.

I wish you the best of luck in all of your examinations, and remember that you can be a top geographer!

The Geography Common Entrance exam

The Geography Common Entrance exam is one hour long. The paper is split into three sections: location knowledge, Ordnance Survey and thematic studies. You should answer all the questions in all three sections.

The location knowledge and Ordnance Survey sections are each worth 10–15 marks and the thematic studies section is worth 50–60 marks. Another 20 marks come from your fieldwork investigation. This makes a total of 100 marks or 100 per cent. Your Common Entrance paper will be marked by the senior school that you hope to attend. The school will work out your final percentage and turn this into a grade (A, B, C, etc.). The percentage required to obtain a particular grade differs between schools.

Location knowledge

- It is best to start with this section of the exam as it can be completed quickly and easily if you have learnt your locations.
- You should spend about **eight minutes** on this section.
- Make sure that you read the questions carefully. If, for instance, you are asked for the name of a country do *not* write the name of a city!
- If you are asked to mark something on a map, such as a line of latitude or a mountainous area, do not forget to label it.
- Make sure that you practise marking the locations on the continent and world maps in this book.
- This is the most straightforward section to revise for as it is just a case of learning and practising. Quizzes with your family and friends will also help you to revise.

Ordnance Survey

- Ensure that you have a sharp pencil, a ruler and a scrap of paper or a piece of string.
- It is important that you have a flat surface onto which you can place the OS map. You may need to move some items from your desk onto the floor.
- You should spend **10–12 minutes** on this section.
- Make sure that you read the instructions carefully and double check all your answers. If the question asks for a distance, do *not* give a direction as your answer!
- Ensure that you always add the correct units to any answer. Use kilometres (km) for distance and metres (m) for altitude.
- Give a six-figure reference for any spot (small) features such as a post office or milestone but a four-figure reference for large features such as woodlands or towns.
- Ensure that you look carefully at the word 'from' in a direction question so that you do not 'go' the wrong way.
- If you are asked to describe a route, remember to break the route into sections and give altitudes, directions and distances, and mention any features that you pass.

Thematic studies

- You should spend between **35–40 minutes** on this section.
- There will be questions from each of the five themes: earthquakes and volcanoes, weather and climate, rivers and coasts, population and settlement, and transport and industry.
- Some questions may refer back to the OS map; other questions may use resources such as photos, graphs or diagrams. You must study these carefully before answering the question. (Remember that the line on a climate graph is the temperature and the blocks are the rainfall.)
- You will be given marks for including examples and for drawing relevant diagrams, even if the question does not specifically ask you to do this.
- If you are asked a question about a case study, make sure that you make your answer specific by using names of places and actual figures.

General points

- Have a watch on your desk. Work out how much time you need to allocate to each question and try to stick to it.
- Make sure you read and understand the instructions and rules on the front of the exam paper.
- Always read the questions carefully, underlining, circling or highlighting key words or phrases.
- Look at the number of marks available in order to assess how much to write for each answer. If you use bullet points to answer a question that offers a high mark, you must make sure that the bullet points include sufficient detail.
- Do not leave blanks. If you do not know the answer, take an educated guess. Wrong answers do not lose marks.

- Make sure that all your diagrams are clearly annotated (labelled with explanations). There are certain diagrams that it is essential you know how to draw. These are clearly marked throughout this book.
- Whenever possible, include impressive geographical terms from the lists of 'word you need to know'. This creates a good impression and will gain you higher marks.
- If a question is particularly hard move on to the next one. Go back to it if you have time at the end.
- Organise your time so that you have time to check your answers at the end.

Command words

Make sure you completely understand these words and phrases. Cover up the definitions with a sheet of paper in order to test yourself.

annotate add descriptive explanatory labels
choose select carefully from a number of alternatives
complete finish, make whole
define give an exact description of
describe write down the nature of the feature
develop expand upon an idea
explain write in detail how something has come into being and/or changed
give show evidence of
identify find evidence of
list put a number of examples in sequence
mark and name show the exact location of and add the name
name give a precise example of
select pick out as most suitable or best
shade and name fill in the area of a feature and add the name
state express fully and clearly in words
study look at and/or read carefully
suggest propose reasons or ideas for something

These words are only used in the scholarship exam:

discuss present viewpoints from various aspects of a subject
elaborate similar to expand and illustrate
expand develop an argument and/or present greater detail on
illustrate use examples to develop an argument or theme

Tips on revising

Get the best out of your brain

- Give your brain plenty of oxygen by exercising. If you feel fit and well, you will be able to revise effectively.
- Eat healthy food while you are revising. Your brain works better when you give it good fuel.
- Think positively. Give your brain positive messages so that it will want to study.
- Keep calm. If your brain is stressed it will not operate effectively.

- Take regular breaks during your study time.
- Get enough sleep. Your brain will carry on sorting out what you have revised while you sleep.

Get the most from your revision

- Don't work for hours without a break. Revise for 20–30 minutes then take a 5-minute break.
- Do good things in your breaks: listen to your favourite music, eat healthy food, drink some water, do some exercise, juggle. Don't read a book, watch TV or play on the computer; it will conflict with what your brain is trying to learn.
- When you go back to your revision, review what you have just learnt.
- Regularly review the facts you have learnt.
- Use past papers to familiarise yourself with the format of the exam.

Get motivated

- Set yourself some goals and promise yourself a treat when the exams are over.
- Make the most of all the expertise and talent available to you at school and at home. If you don't understand something, ask your teacher to explain.
- Find a quiet place to revise and make sure you have all the equipment you need.
- Organise your time so that you revise all subjects equally.

Tips on taking the exam

Before the exam

- Have all your equipment and pens ready the night before. You will need: ruler, calculator, red, yellow and blue colouring pencils, two normal pencils, an ink pen and spare cartridges.
- Make sure you are at your best by getting a good night's sleep before the exam.
- Have a good breakfast in the morning.
- Take some water into the exam if you are allowed.
- Think positively and keep calm.

1 Earthquakes and volcanoes

1.1 The Earth's structure, tectonic plates and plate boundaries

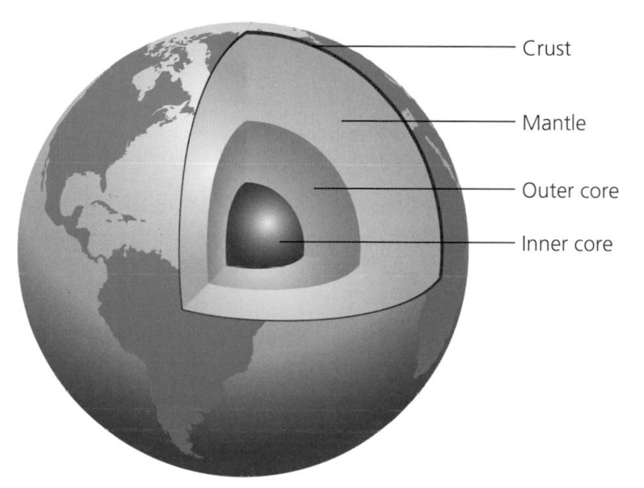

■ Figure 1.1: The structure of the Earth

The Earth is made up of the following parts:

- Crust – the solid outer layer of the Earth on which we walk
- Mantle – the next layer towards the centre of the Earth. This is the thickest layer and is made of semi-molten rock called magma.
- Outer core – the next layer towards the centre of the Earth. This is made of liquid metal.
- Inner core – the centre of the Earth. This is made of solid metal and is at a temperature of up to 5500 °C.

Tectonic plates are the huge slabs of rock that form the Earth's crust and that float on the mantle (the semi-solid rock beneath the crust).

- Continental plates are thick but light in weight (less dense) and form land. They are made of granite.
- Oceanic plates are thinner but heavier (more dense) and have sea over them. They are made of basalt.

The movement of plates is called continental drift. This can push the plates together or push them apart. Continental drift occurs due to the movement of the magma in the mantle below the plates. The movement of the magma is caused by convection currents generated by the immense heat at the Earth's core.

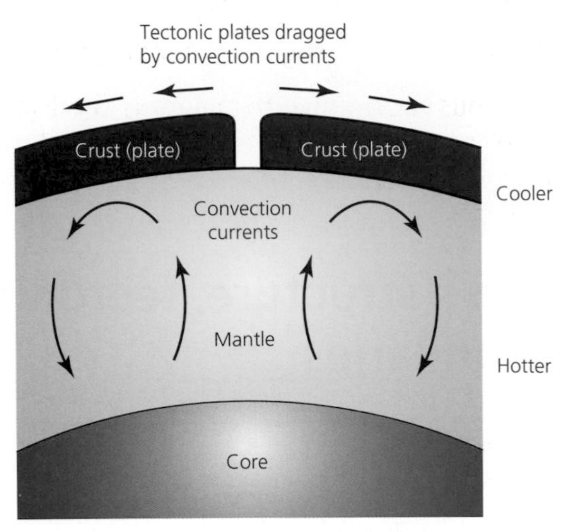

■ Figure 1.2: Continental drift and plate movement

The edges where the plates meet are called plate boundaries. There are four types of plate boundary: constructive, destructive, conservative (or sliding) and collision.

Constructive plate boundary

- At a constructive plate boundary two plates move apart.
- Magma rises to the surface, due to gas bubbles in the magma that make it lighter than the surrounding rock.
- Volcanoes are formed.
- Gentle eruptions occur which may continue for years.

Most constructive boundaries are under the sea and form chains of volcanic islands. The Mid-Atlantic Ridge is the most famous of these chains.

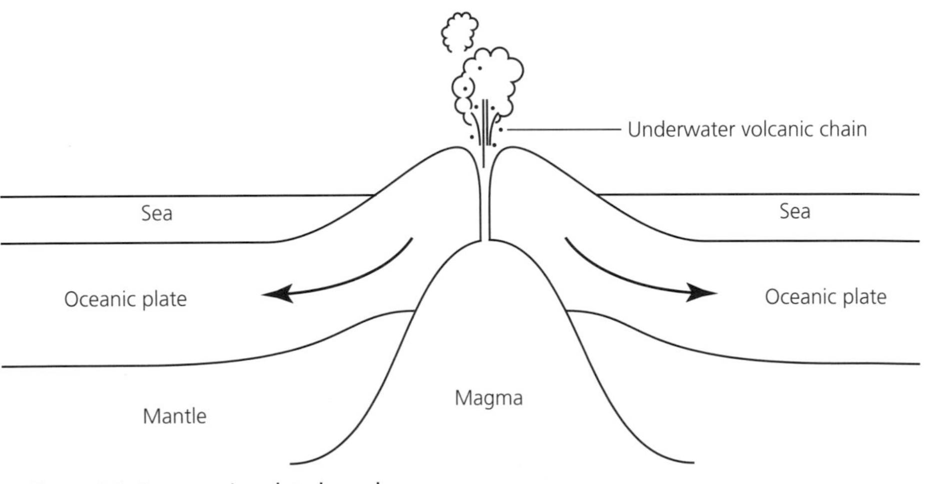

■ Figure 1.3: Constructive plate boundary

Destructive plate boundary

- At a destructive plate boundary, an oceanic and a continental plate collide.
- The heavier oceanic plate sinks under the continental plate into what is known as a subduction zone.

- The melted crust rises (due to the gas bubbles in the magma that make it lighter than the surrounding rock) to form explosive, dangerous volcanoes.
- When the two plates rub together, friction occurs, leading to earthquakes.

The most famous destructive boundary is the Pacific Ring of Fire, which forms a band of earthquakes and volcanoes round the edge of the Pacific Ocean. A destructive plate boundary was also the cause of the Soufrière Hills volcano in Montserrat.

You need to know how to draw this diagram.

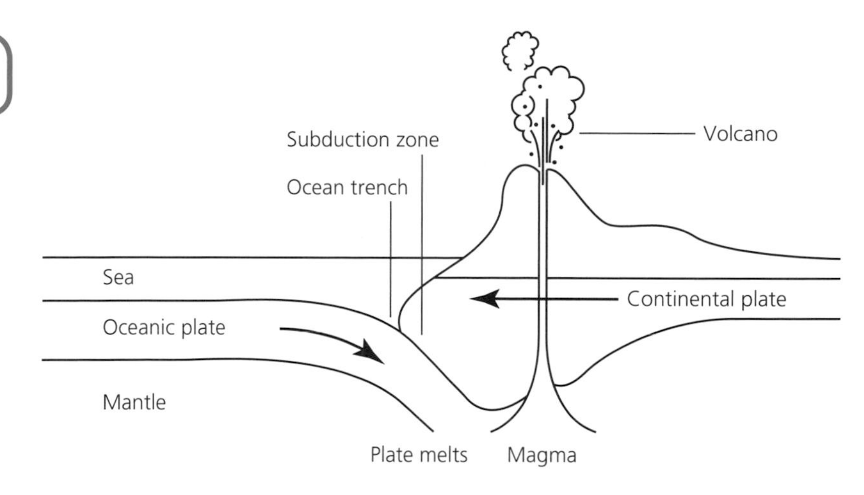

■ Figure 1.4: Destructive plate boundary

Conservative (or sliding) plate boundary

- At a conservative plate boundary, two plates slide past each other.
- The plates become locked and tension builds up over years.
- Eventually the plates will jolt past each other, causing powerful earthquakes.
- Volcanic activity does not occur.

The most famous of these is the San Andreas fault.

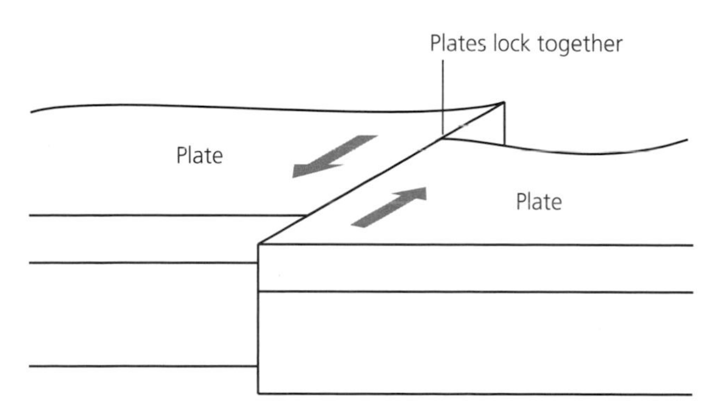

■ Figure 1.5: Conservative plate boundary

Collision plate boundary

- At a collision plate boundary, two continental plates push together.
- Neither sinks beneath the other as they are both made from light rock.
- The plates buckle to form fold mountains and violent earthquakes occur.
- Volcanic activity does not occur.
- The area where the earthquake starts underground is known as the focus. Directly above the focus, on the Earth's surface, is the epicentre.

The Himalayas are the most famous fold mountains.

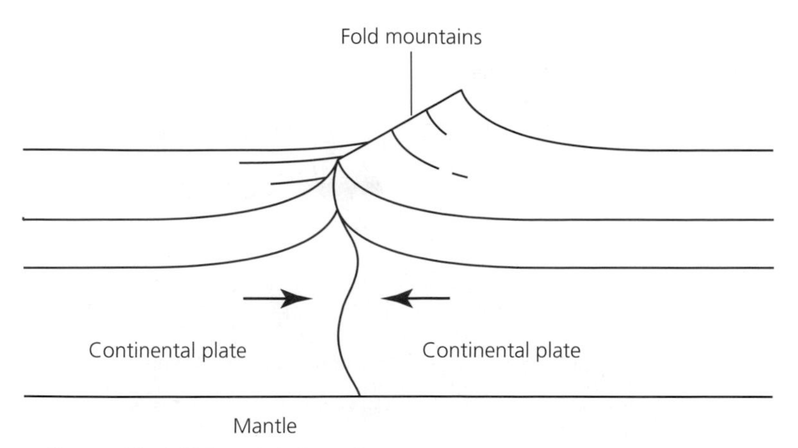

■ Figure 1.6: Collision plate boundary

Revision tip

Try thinking of hand gestures for each of the four plate boundaries. For example, a constructive plate boundary could be represented by the fingers of your two hands touching on their tips and then moving apart, a collision plate boundary could be represented by clapping your hands. You could make sure that in each gesture, each flat hand represents a plate. Now test your friends.

1.2 Types of volcano

There are two types of volcano: composite and shield.

Composite volcanoes

- These occur in areas of destructive plate boundaries.
- The eruptions are violent, ejecting thick and sticky lava.
- Ash and lava are ejected into the air and descend as slow-flowing, thick lava. The process is then repeated, building up layers of ash and lava.
- Pyroclastic flows (hot gas and ash) travelling more than 160 km per hour can flatten and burn everything in their path.
- Lahars (melted ice or rain mixed with ash) can occur.
- Thick layers of ash leave areas uninhabitable.

Examples of composite volcanoes are the Soufrière Hills, Montserrat and Mount Etna.

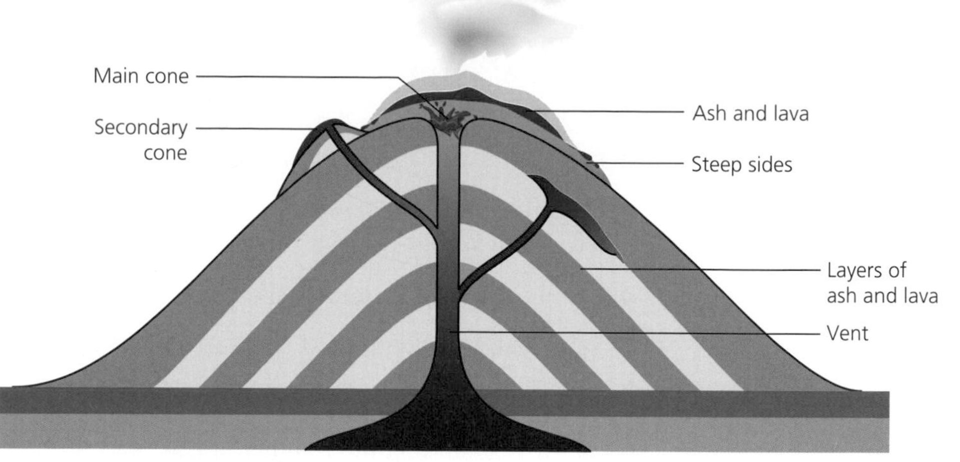

■ Figure 1.7: Composite volcano

Shield volcanoes

- These occur in areas of constructive plate boundaries.
- Wide, gently-sloping volcanoes eject thin, runny lava.
- The eruptions are not explosive and are less likely to result in loss of life.

  Examples of shield volcanoes are Iceland and Hawaii.

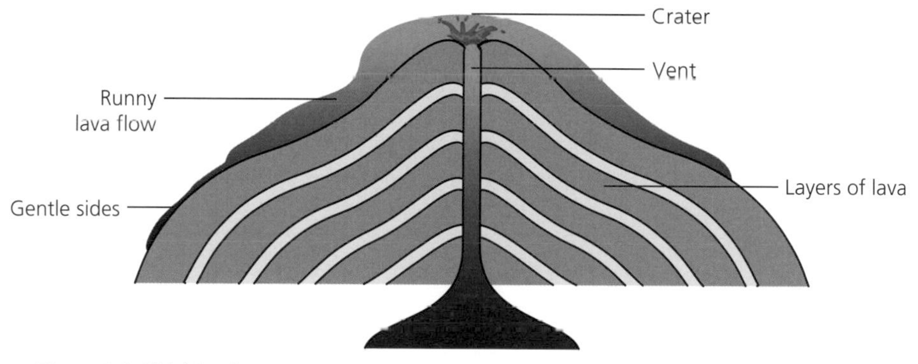

■ Figure 1.8: Shield volcano

1.3 Immediate effects of a volcanic eruption

- Ash fall
- Mud flow (lahar) – ash mixed with rain or melted ice
- Pyroclastic flow – hot gas and ash rolling down the cone
- Lava flow – molten rock
- Pyroclastic bombs (volcanic bombs) – lava cooling when ejected into the air, falling as solid rock

1.4 Preparing for and reacting to a volcanic eruption

- Hazard maps can be drawn (as in Montserrat) to show which areas are safest and which are most at risk.
- Lava flows can be diverted by channels or explosives, dammed or sprayed with cool water.
- People can be evacuated.
- Seismometers can record the earthquakes that occur as the magma rises.
- Tiltmeters can record changes in the shape of a volcano before an eruption.
- Satellites can record changes in the temperature and shape of a volcano before an eruption.

1.5 Why people live near volcanoes

Despite the dangers of living near a volcano, people continue to live in these areas for a number of reasons:

- Interest in volcanoes generates tourism and therefore boosts the local economy.
- Geothermal energy can be produced from the rising steam, for example, in Iceland and New Zealand.
- Fertile soil is produced by the weathering of volcanic ash. This soil is particularly good for grapevines.
- Minerals, such as gold and diamonds, can be found in the area.

Revision tip

You could draw a revision picture to represent each of these facts.

1.6 Causes and effects of earthquakes

Causes

Earthquakes mainly happen on or near to plate boundaries.

- As the plates slide under, over or past each other, friction causes the plates to lock together.
- As the pressure increases, weaknesses or fault lines in the Earth's crust close to the boundaries begin to fail.
- The fault line breaks and the stored energy is released and travels outwards from the focus as seismic waves.

Effects

- The epicentre of the earthquake is on the Earth's surface directly above the focus and is likely to suffer the greatest amount of damage.
- The closer the focus is to the crust the greater the damage.

- The earthquake's energy is recorded by a seismometer, which measures the energy released by the earthquake using the Richter scale (1–10).
- The Mercalli scale (I–XII) is a scale used to measure the damage caused by an earthquake.
- Primary effects happen immediately; they include the destruction of buildings, breaking glass and falling masonry.
- Secondary effects occur hours or days later; these include tsunamis, disease from contaminated water, loss of communications, fire and a damaged economy.

1.7 Preparing for earthquakes

It is very hard to predict an earthquake; only vague predictions can be made from looking at historical patterns of eruptions. However, preparation can be made in areas that are prone to these natural disasters.

- Earthquake drills can be practised in offices and schools.
- Buildings can be built with counterbalances and rubber foundations to withstand even powerful earthquakes.
- Computers can cut off gas supplies as soon as an earthquake breaks, to minimise fires.
- Tsunami walls and shelters can be built in areas prone to this kind of disaster.
- Families can keep survival kits in their homes.

1.8 Factors determining the severity of damage

A number of factors determine the severity of damage caused by a volcano or an earthquake:

- The type of plate boundary that has caused the volcano – destructive plate boundaries cause violent volcanoes.
- The proximity of a volcano or an earthquake's epicentre to a large settlement – those situated near large cities where population is dense cause more deaths than those in less populated areas.
- The proximity of an earthquake's focus to the Earth's surface – the closer the focus the more powerful the earthquake.
- The wealth of the country in which it erupts – a developed country can afford scientific prediction instruments, buildings that are designed to withstand earthquakes, a quick reaction force and good medical care for the injured.
- The time of day when the earthquake strikes – if it strikes when people are congregated in one area, for example, at rush hour, its results can be more devastating.

1.9 Comparing an earthquake or volcano in a developing country with one in a developed country

- The level of death and injury may be greater in a developing country as the hospitals and emergency services are less effective.
- The cost of repair may be greater in a developed country as the infrastructure is more developed.
- More death and destruction may occur around a volcano in a developing country as many subsistence farmers will farm close to the volcanic cone in order to benefit from the fertile soil.
- The amount of aid received is probably going to be greater in a developing country as the population's needs are greater.
- Greater scientific monitoring and data gathering will occur in developed countries. Therefore prediction will be more accurate in developed countries, although predicting an earthquake is very difficult.
- Emergency action plans are less likely to be prepared or practised in developing countries.
- Secondary effects may be worse in a developing country, as the level of poverty means that disease is more likely to spread.

Revision tip

You could make a mind map to revise the whole volcanoes and earthquakes topic. Write 'volcanoes and earthquakes' in the centre and then add sticks coming off for the structure of the Earth, plate boundaries, types of volcano, immediate effects of a volcanic eruption, why people live near volcanoes, causes and effects of earthquakes, preparing for earthquakes, factors determining the severity of damage and how effects differ between a developed and developing country. Each of these categories could then have sticks coming off them explaining all of the important points. Remember to use memory pictures rather than words to represent the facts. Use lots of colours but try to use colour to good effect.

1.10 Case study – Soufrière Hills volcano, Montserrat, 1995 to present

Facts

- After a long period of dormancy the Soufrière Hills volcano became active in 1995 and eruptions have continued to the present day.

Cause

- The Soufrière Hills volcano is located on a destructive plate boundary of three plates: South American, North American and Caribbean.

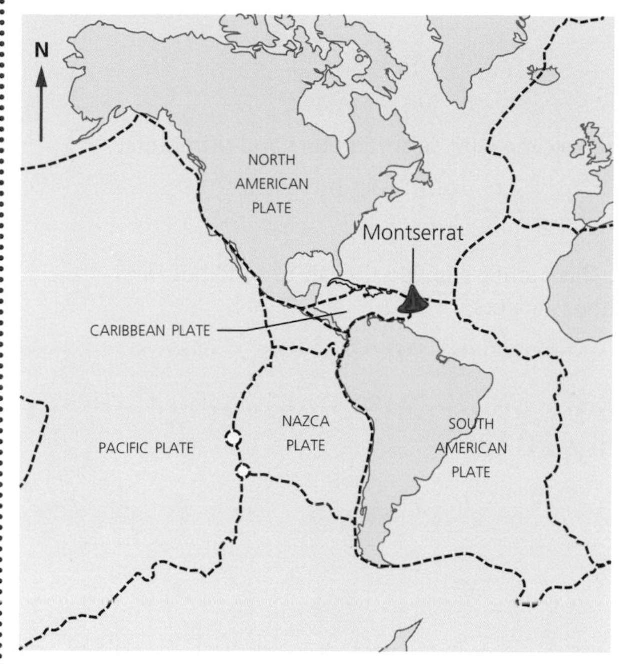

■ Figure 1.9: Location of Soufrière Hills volcano

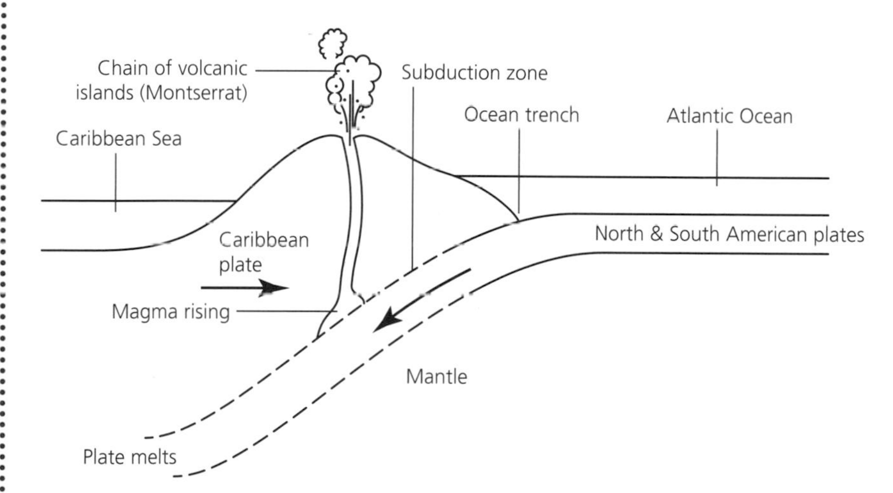

■ Figure 1.10: Destructive plate boundary (North and South American and Caribbean plates)

Effects

Environmental

- Pyroclastic flow burned vegetation.
- Ash covered two-thirds of the island.
- Wildlife disappeared due to ash fall and pyroclastic flow.
- Coral reef and sea creatures died from the ash washed into the sea.
- New land was formed when the pyroclastic flow solidified in the sea.

Economic

- The agricultural economy was ruined.
- The tourist economy was ruined – the airport was closed for ten years.

Social

- Montserrat is a UK dependency therefore the UK is obliged to offer aid.
- 60 per cent of housing was destroyed.
- 23 people died in the pyroclastic flow of 25 June 1997.
- 8000 inhabitants left Montserrat for the UK and Antigua.
- Hospitals were destroyed.
- Few schools were left intact.
- There was a lack of clean water and sewerage facilities.

Human response

- The Montserrat volcano observatory was built to monitor the volcano with seismometers and tiltmeters.
- A hazard map was drawn up to highlight danger and safe zones.
- International aid was received from the UK and other countries.
- Inhabitants were evacuated to the UK and Antigua.
- There was a concert in the Royal Albert Hall to raise money. This money allowed the UK to send out *HMS Liverpool* with emergency showers and kitchen facilities for the islanders.

1.11 Case study – Mount Etna volcano, Sicily, 2013

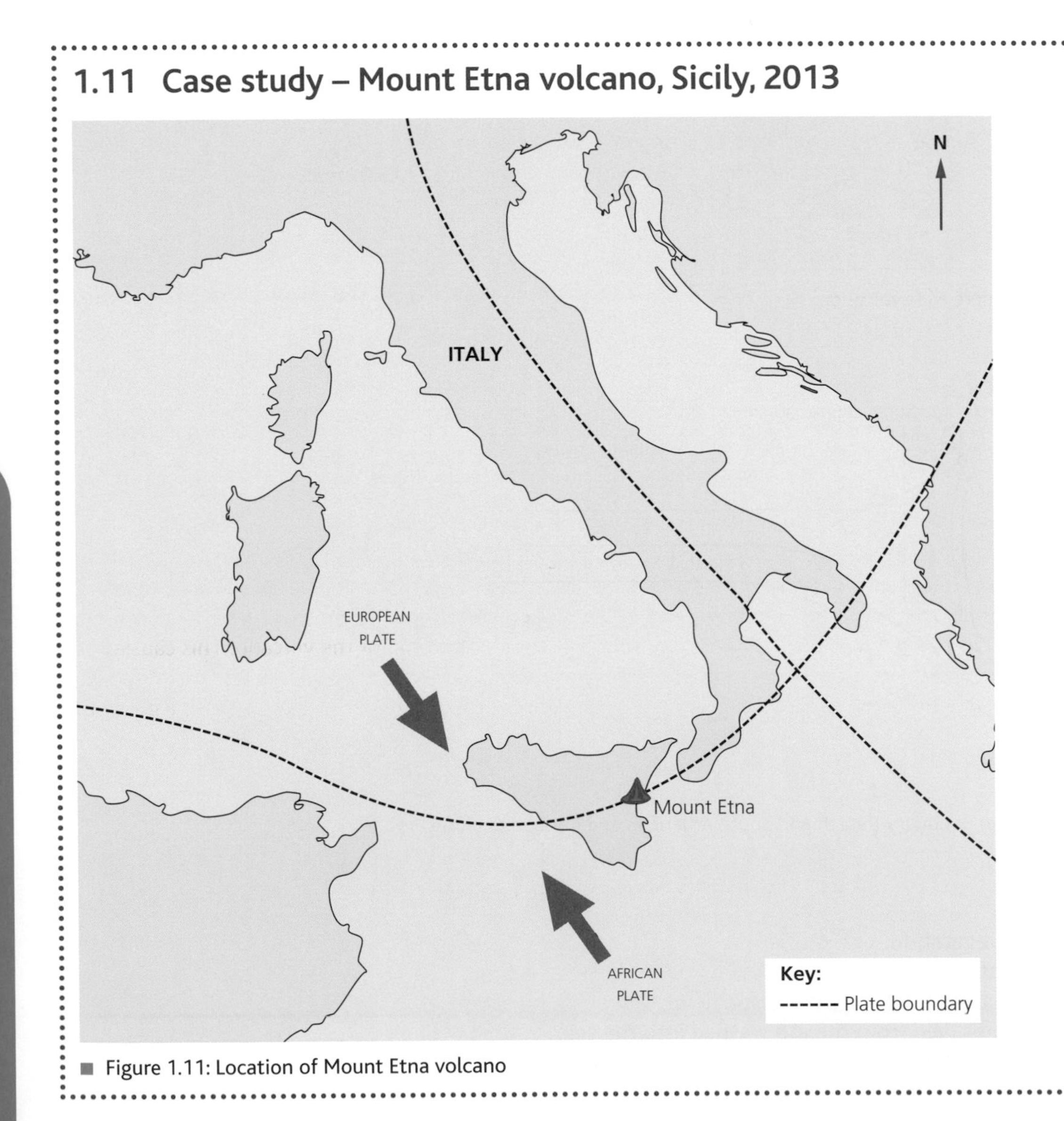

■ Figure 1.11: Location of Mount Etna volcano

Facts

- Mount Etna erupted thirteen times in 2013.
- It is the tallest volcano in Europe, measuring 3329 m.
- It is constantly spewing ash, smoke and lava from its many craters.
- It is a composite volcano.

Cause

- Mount Etna is located on a destructive plate boundary where the African plate subducts under the Eurasian plate.

Effects

Environmental

- The lava flows reached temperatures of 1000 °C.
- Millions of tonnes of water vapour and 50 000 tonnes of carbon dioxide were released into the atmosphere each day.
- Smoke and ash rose 900 m into the air.
- A layer of black ash covered cars 50 km away.

Economic

- Nicolosi, a village at the base of Etna, has been repeatedly destroyed by lava.
- Less snow settles on the sides of the volcano, which affects the ski season.
- Catania airport, which is the main airport in Sicily, was forced to close while the runways were cleared of ash.
- The agricultural and tourist economies of the towns on Etna's slopes were affected.

Human response

- Mount Etna is the most studied volcano.
- Mayors from towns surrounding the volcano called for urgent government action.
- The authorities stopped access to the volcano and strict regulations were put in force to protect people. A hazard map was drawn up to show areas that were out of bounds.
- In the past, dams of earth and volcanic rock were put up to protect the tourist areas. These diverted the flow and kept it under control.
- In the past, the Italian government has promised to reduce taxes for villagers to help them get through the crises, and handed out more than £5.6 million in aid.

1.12 Case study – Eyjafjallajökull volcano, Iceland, 2010

Facts

- The volcano is located under the Eyjafjallajökull glacier in the southern part of Iceland.
- The Eyjafjallajökull volcano began erupting in March 2010. There was an explosive phase from 14 to 21 April, followed by a drop in intensity and then increased activity at the start of May.
- As the lava came out of the volcano it cooled very quickly as the glacier was on top of the volcano. This caused the lava to shatter into tiny fragments of ash.
- The glacier melt water poured into the vent, which created steam and caused the mineral ash to reach heights of 6–10 km.

Causes

- Iceland is situated on the Mid-Atlantic Ridge, a constructive plate boundary running through the middle of the Atlantic Ocean.
- The Eurasian plate is moving very slowly eastwards and the North American plate is moving very slowly westwards. The two plates are diverging.
- The Eyjafjallajökull glacier lies on top of the volcano.

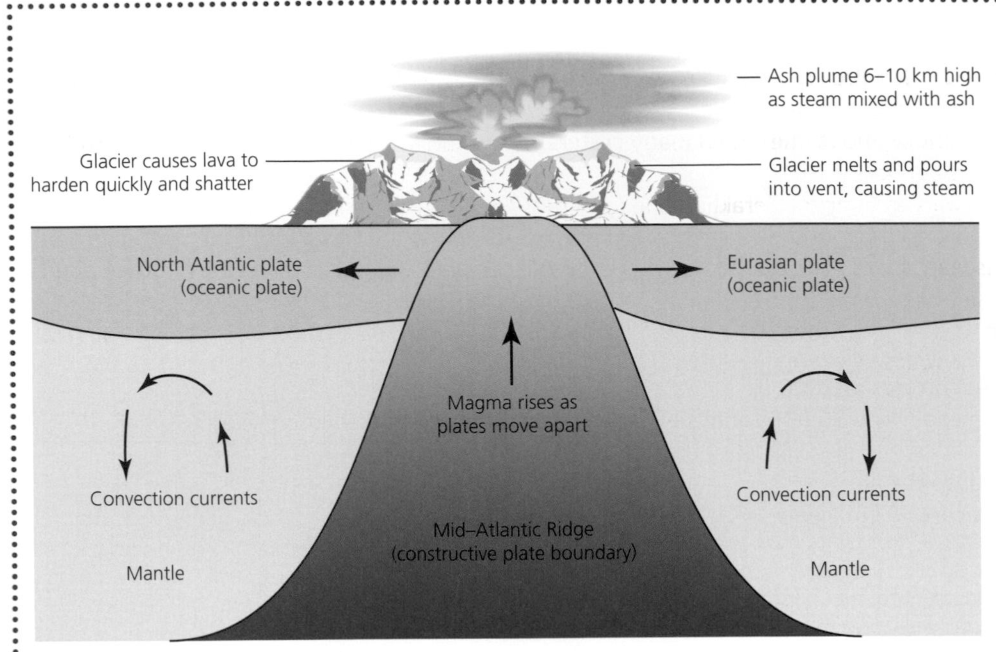

Figure 1.12: Cause of Iceland volcano

Effects

Environmental

- The volcano released lots of greenhouse gases which are harmful to the atmosphere.
- Some say the volcano benefited the environment as 95 000 flights did not take place as a result of the eruption.
- Scientists say that the volcano may just be awakening, that its previous eruption lasted for two years and that they are expecting many future eruptions. They say this eruption may spark the eruption of its neighbour, Mount Katla, a bigger and more powerful volcano that would cause more havoc.
- Large amounts of volcanic ash fell onto the town of Vik in Iceland.
- Areas around the volcano were flooded by the melting glacier.
- Land near to the volcano was covered in ash and this poisoned animals.

Economic

- Eurocontrol traffic control closed much of the airspace in Europe because of the risk of ash harming aeroplane engines. Ash clouds pose a danger to aircraft and can lead to engine failure. The fine, abrasive particles erode metal, clog fuel and cooling systems and melt to form glassy deposits. Flight instruments, windows, lights, wings and cabin air supply can also be affected. However, airspace was closed based on theoretical models, not on facts.
- After being closed for six days, British airspace reopened.
- 95 000 flights were cancelled and it took several weeks to clear the backlog of flights.
- Global airlines lost about £1.1 billion of revenue.
- Tens of thousands of people were stranded, unable to get to work. Businesses around the world were affected.
- Producers of perishable goods, such as food and flowers, were hit hard as their goods were left at airports.
- Car maker Nissan had to suspend production for a day as it could not import components.

Social

- The volcano brought European air traffic to a standstill.
- 1.2 million passengers a day could not fly.
- Many people missed weddings, funerals or long-awaited holidays because of the halt in air transport.
- Many people who lived under flight paths enjoyed the lack of noise for a few days.
- Scientists said that the ash was travelling high in the atmosphere, was not likely to come down and if it did come down, it would be too diluted to have any affect on people in the UK.
- The World Health Organization said that falling ash was 'more bothersome than hazardous to your health'.
- Doctors advised patients with asthma, bronchitis, emphysema or heart disease (those most at risk) to remain indoors to avoid irritation to their throat and lungs, a runny nose or itchy eyes.
- Icelandic farms near to the volcano were damaged. The ash fall poisoned animals and destroyed farmers' livelihoods.

Human response

- The then UK Prime Minister, Gordon Brown, ordered three Navy ships to be sent to help stranded Britons unable to return home.
- Ryanair laid on extra flights from Spain to the UK and has now had to comply with EU rules and pay for stranded passengers' food and accommodation.
- Thomas Cook sent rescue planes to Cancun, Heraklion and Sharm-el-Shake and returned 2500 passengers.
- Insurance companies refused to pay travellers extra accommodation costs or the cost of alternative journeys home as they said that this was an act of God and therefore exempt from payment. In the end most airlines paid out reasonable costs to holidaymakers.
- The emergency services in Iceland evacuated hundreds of people from the area before any water flowed downhill and dug trenches through the roads to allow flood waters to pass without washing away bridges.

1.13 Case study – L'Aquila earthquake, Italy, 2009

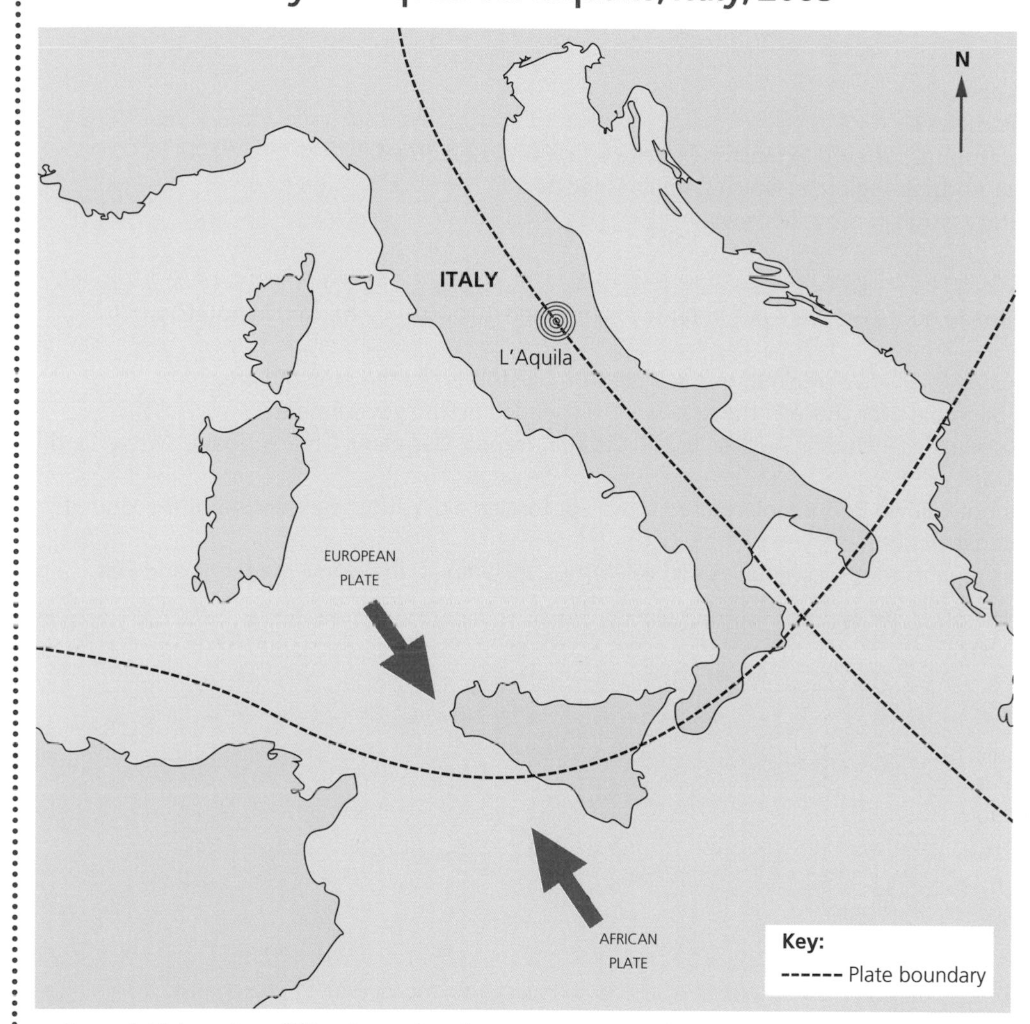

■ Figure 1.13: Location of L'Aquila earthquake

Facts

- An earthquake with a magnitude of 6.3 on the Richter scale hit on 6 April 2009.
- There had been around 100 minor tremors since the previous January and two smaller earthquakes the previous day.
- Later an aftershock of 5.3 magnitude occurred.
- The focus was 10 km below the Earth's surface.

Cause

- The Eurasian and African plates meet along a line which runs through North Africa and crosses the Mediterranean near southern Italy and Greece. As a result two main cracks (fault lines) cut across the Italian peninsula, one running north–south along the spine of the Apennine mountains and another crossing east–west south of Rome and north of Naples.
- The region surrounding L'Aquila is therefore criss-crossed by fault lines.
- The city is built on the bed of an ancient lake, which exaggerates the seismic waves.
- To the east in the Adriatic Sea, the Earth's crust (a mini plate called the Adria) is subducting under Italy.

Effects

Economic

- The earthquake cost Italy $4 billion.
- Tourism in the historical city declined.
- 26 cities and towns were damaged. Poor building standards and construction materials meant that buildings collapsed easily.
- 3000 to 11 000 buildings in the medieval city were damaged. The dome on the church of St Augustine collapsed and damage was caused to the city archives.

Social

- 308 people are known to have died.
- 28 000 people were left homeless.
- The new wing of L'Aquila Hospital suffered extensive damage and was closed down.
- Schools were closed in the Abruzzo region, in which L'Aquila is located.
- The aftershock caused safety issues for rescue crews.

Human response

- 40 000 people who were made homeless were put in tented camps and 10 000 were housed in hotels on the coast.
- All Italian mobile phone companies gave free minutes and credit to customers in the affected area.
- Tax billing and mortgage payments for those in the area were suspended by the government.
- Aid was offered by many countries including Austria, Brazil, Croatia, France, Germany, Greece, Spain, Switzerland, Tunisia, Ukraine and the USA.
- The Prime Minister at the time, Silvio Berlusconi, refused to accept foreign aid, with the exception of the United States' offer of aid for reconstruction.
- Aid was offered by various organisations, companies and celebrities, including Carla Bruni, Madonna and Fiat. This was also declined.

1.14 Case study – Haiti earthquake, 2010

Facts

- The quake struck on 12 January 2010 at 4.53 p.m.
- The quake measured 7 on the Richter scale.
- The epicentre was 15 km south-west of Port-au-Prince.
- The earthquake was quickly followed by two strong aftershocks of 5.9 and 5.5 magnitude.
- Haiti forms part of the island of Hispaniola. Hispaniola contains two countries: Haiti and the Dominican Republic.

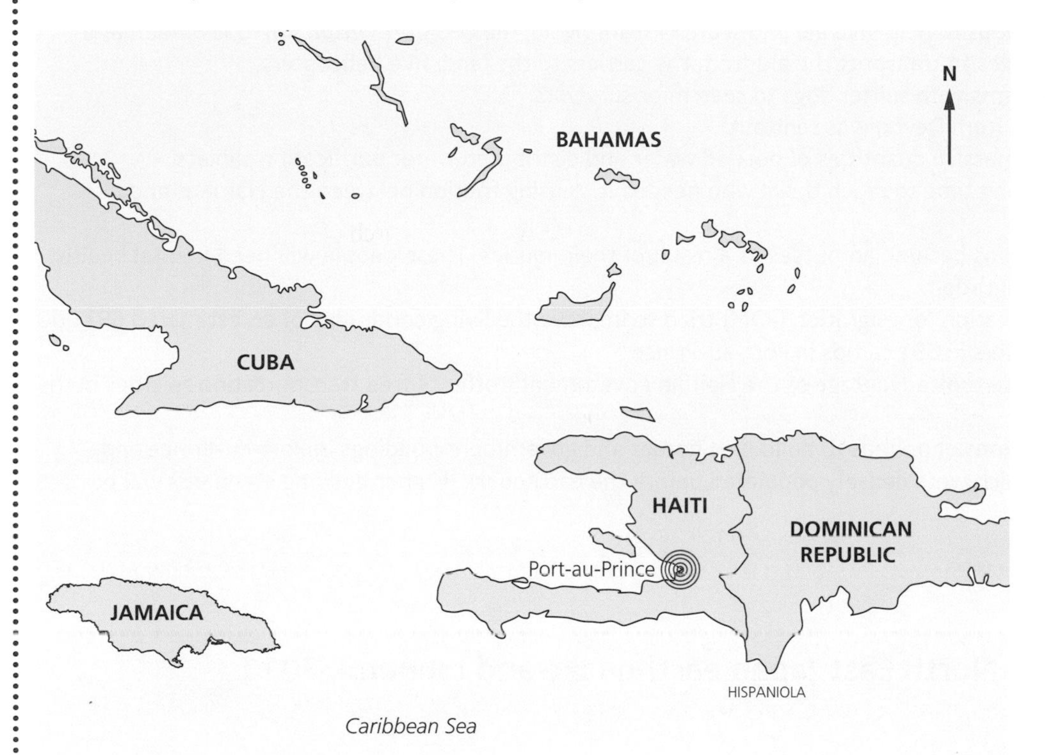

Figure 1.14: Location of Haiti earthquake

Causes

- The conservative plate boundary between the Caribbean and North American plates runs right through Haiti.
- These two plates constantly slide past one another, at about 2 cm per year, with the Caribbean plate moving eastward with respect to the North American plate.
- The focus of the Haiti quake was 10 km below the Earth's surface. This is very shallow, so the energy that was released was very close to the surface and caused violent ground shaking.

Effects

Social

- Haiti's government says about 230 000 people died in the earthquake.
- About 300 000 were injured.
- The survivors had nowhere to go as the hospitals were full.
- Patients were treated without proper doctors and medical equipment as the doctors had been killed or injured and the equipment had been destroyed.
- There was a lack of food meaning that even those who survived suffered extreme hunger.
- There was an acute shortage of drinking water. The water supply system, which before the disaster only provided 40 per cent of the population of Port-au-Prince with clean water, had effectively collapsed.
- Mass graves were commissioned, holding up to 7000 corpses each. The remaining bodies, which could not be buried, were used as a road block to protest against the lack of aid.
- As people tried to help their families and others, violence and looting broke out, which could not be controlled due to the lack of police.
- The high death rate was due to a number of factors: 72.1 per cent of the population lived on less than $2 a day; the people lived in packed shanty towns in poorly constructed buildings; builders did not follow safety codes and bribes were given so that builders could take short cuts; the quake hit close to a poorly constructed, large urban area.

Economic

- The infrastructure of the capital city Port-au-Prince was completely ruined. The roads were impassable, the country's tiny airport was full of aid planes and the port was full of rubbish.
- The Presidential Palace and all other ministries collapsed, leaving parliament without a base from which to make decisions.
- Almost every house in the country fell down due to their unstable foundations.

Human response

- The UN said that £165 million was pledged (£10 million from the UK and a further £10 million from the UN).
- The USA sent: 10 000 troops to help give aid and search for survivors; the *USS Carl Vinson, the USS Bataan* and several amphibious vehicles to transport the aid from the carriers to the land; five helicopters.
- Indonesia sent search teams with sniffer dogs to search for survivors.
- A mountain rescue team from Devon was sent out.
- Aid agencies shipped in massive quantities of bottled water and distributed water purification tablets.
- Most of this aid took a long time to reach those who needed it, causing friction between the Haitians and aid workers.
- Many thousands of Haitians became amputees as a result of their injuries. These people will need mental health services as well as rehabilitation.
- The International Organization for Migration (IOM) tried to improve the living conditions of an estimated 692 000 people in makeshift shelters in 591 camps in Port-au-Prince.
- More than 235 000 people took advantage of the Haitian government's offer of free transportation to cities in the north and south-west.
- Haiti is likely to need billions of pounds to build new homes and government buildings in Port-au-Prince and the surrounding area, which were densely populated before the earthquake. Higher building standards will be necessary.

1.15 Case study – North East Japan earthquake and tsunami, 2011

Facts

- The quake struck on 11 March 2011.
- The earthquake measured 9 on the Richter scale.
- The epicentre was 130 km from the coast of Sendai in North East Japan.

Causes

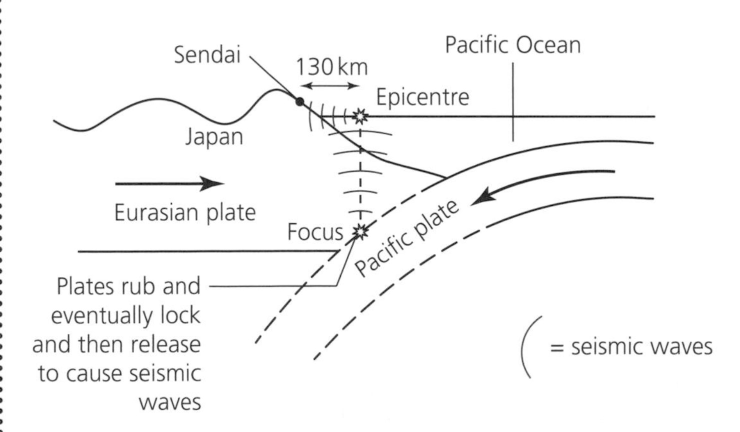

Figure 1.15: Cause of Japanese earthquake and tsunami

- Japan lies on many plate boundaries.
- This earthquake was caused by a destructive plate boundary: the Pacific plate was sinking under the Eurasian plate.
- This plate boundary is known as the Pacific Ring of Fire, an extremely active tectonic area.
- The ocean trench and the subduction zone were in the Pacific Ocean.
- As the Pacific plate slipped under it released an enormous amount of tension at the focus. This had been building up over centuries and caused massive primary seismic waves.

- The more dangerous secondary seismic waves made their way towards the eastern side of Japan.
- The shock waves also caused a tsunami.
- There were 508 aftershocks of magnitude 5, 6 and 7.

Effects

Environmental

- The earthquake shifted the Earth's axis by 25 cm.
- The Earth has changed shape as a result of the earthquake.
- Japan's north-east coast moved 3 metres out to sea.
- Parts of the coast dropped over 1 metre, causing tsunami defence walls to become smaller.
- The earthquake ripped down the infrastructure in the area affected.
- The Fukushima nuclear power plant was affected by both the tsunami and the earthquake. Meltdown occurred.
- Liquefaction appeared in the cracks in the ground. Liquefaction is when water is pushed up to the surface and through the cracks, like a sponge releasing water.
- 10 billion tonnes of water spread out across Japan after the tsunami.
- At Sendai the water travelled 10 km inland.
- The stress released in the 2011 earthquake may trigger a bigger earthquake further south in Japan, in places such as Tokyo.

Social

- 15 000 people died and 5000 people were missing, presumed dead.
- The Fukushima nuclear power plant was affected and sent out radiation. Some people were evacuated from the nearby area.
- The airport at Sendai was destroyed.
- The temporary shelters in Sendai were full.
- Huge areas were without power.
- Services such as schools and hospitals were shut or destroyed for a long period of time.
- Mountain residents worried about the corpses being swept up the rivers in the water.
- Hawaii had to evacuate people living near the sea. Luckily no one died.
- In Minamisanriku 50 per cent of the population died and 95 per cent of the buildings were destroyed.
- People had to rebuild their lives; many people had lost everything.

Human response

- The immediate response to the earthquake was automatic warnings to the Japanese people, on mobile phones and on television.
- There was a one-minute warning for earthquake and a 20-minute warning for the tsunami.
- The Hawaii Pacific Earthquake facility also sent warnings to the countries surrounding the Pacific, and as result only one person died in a country other than Japan (on the California coast).
- More data was collected from this earthquake than from any other disaster to date.
- The Japanese military were immediately on the scene, clearing debris in the towns that had been demolished.
- Aid was sent by different countries to help ease the impact of the disaster. Rescue groups were also sent.
- Every school and business had performed earthquake drills and these saved many lives.
- The Japanese people remained calm and waited for help; there was not a mass panic.

Revision tip

Whichever two case studies you have chosen it is a good idea to make mind maps of the details. Put the name of the case study in the middle with the date on which it occurred and then add a stick each for causes, effects (split into environmental, social and economic) and human responses. Remember to use colour to good effect and to use revision pictures and words rather than sentences.

Exam-style questions

Try these questions, using the Soufrière Hills volcano as your case study. Answers are near the back of the book.

1.1 Locate the case study on a world map. (1)

1.2 Why did this volcano occur? (4)

1.3 For an earthquake or a volcano that you have studied, describe the major effects that it has had on the surrounding area. (4)

1.4 Study the map below which shows the distribution of earthquakes and volcanoes around the world. Name areas A and B. (2)

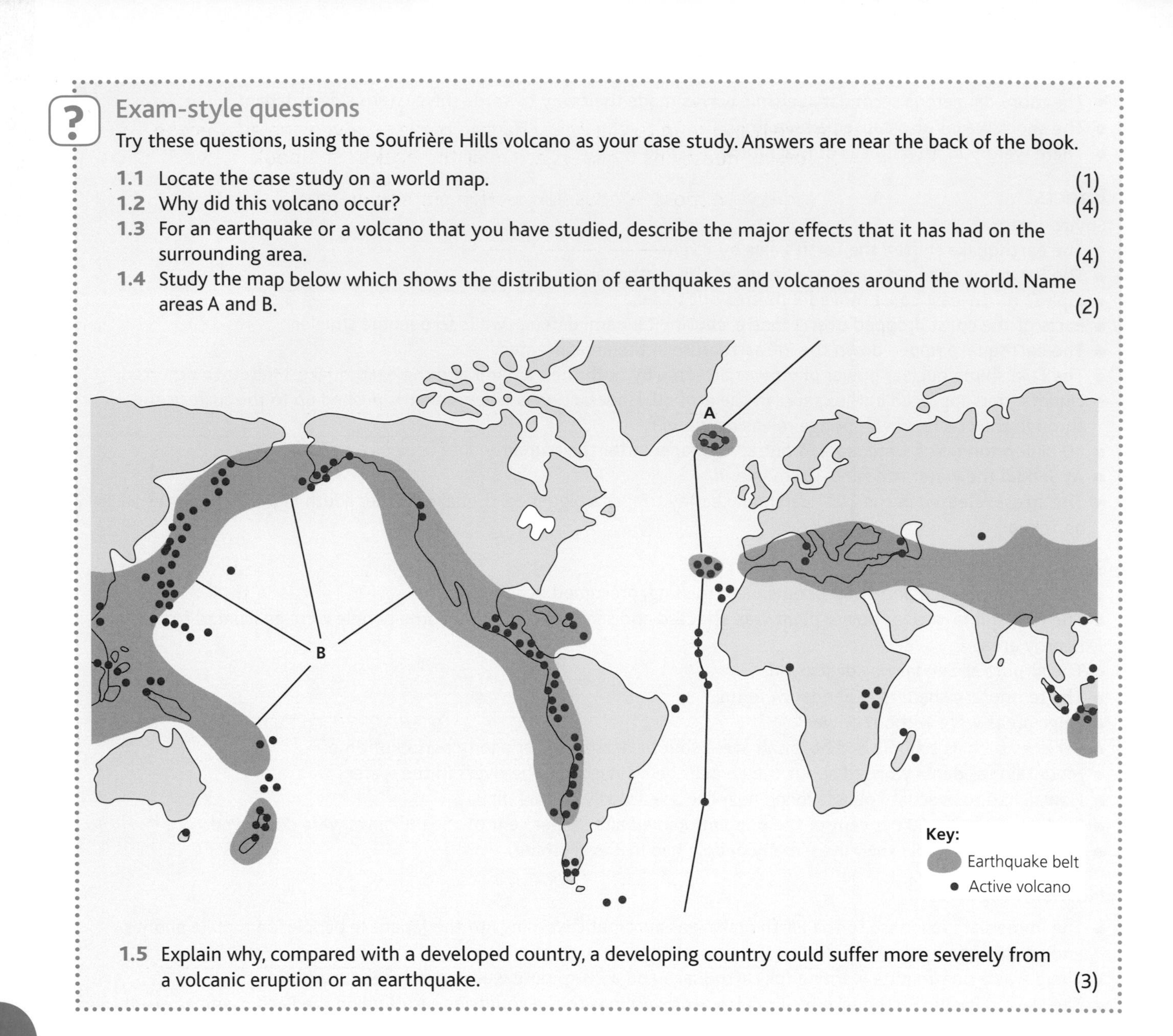

1.5 Explain why, compared with a developed country, a developing country could suffer more severely from a volcanic eruption or an earthquake. (3)

★ Make sure you know

- ★ The different types of plate boundaries
- ★ Why plates move
- ★ How to annotate a destructive plate boundary
- ★ The immediate effects of a volcanic eruption or an earthquake
- ★ What can be done to prepare for a volcanic eruption or an earthquake
- ★ The reasons why people live near volcanoes
- ★ How the effects of earthquakes and volcanoes can differ according to level of development of a country
- ★ The causes, effects and human response to a volcanic eruption and an earthquake (one having occurred in a less developed country and one in a more developed country)

Test yourself

Before moving on to the next chapter, make sure you can answer the following questions. Answers are near the back of the book.

1 Name the most famous destructive plate boundary.

2 Why do plates move?

3 What instrument records the change in shape of a volcano?

4 Write the definitions of these words and phrases you need to know, and then ask someone to check them. (Words and phrases in italics are useful even though you are not required to know them.)

collision boundary

conservative plate boundary

constructive plate boundary

core

crater

crust

destructive plate boundary

dormant

earthquake

epicentre

extinct

focus

fold mountains

foreshock

geothermal energy

plate tectonics

lahars

pyroclastic flow

lava

seismic wave

magma

seismometer

magma chamber

subduction zone

mantle

tectonic plates

Mid-Atlantic Ridge

tsunami

Pacific Ring of Fire

vent

Pangea

volcanic bombs

plate boundary

volcano

Weather and climate

The weather and the climate are two different things:

- **Weather** is the hour-to-hour, day-to-day condition of the atmosphere (wind speed, wind direction, temperature, humidity, sunshine, type of precipitation).
- **Climate** is the average weather conditions for an area over a long period of time. The climate is often shown on a climate graph.

2.1 The water cycle (hydrological cycle)

Within the water cycle, water moves from one state to another. This drives our weather.

1. The water cycle begins when water from the sea or a lake evaporates to form water vapour. Water from plants is also turned to water vapour by transpiration.
2. This water vapour then rises, cools and condenses to form clouds.
3. As the clouds rise further and cool, precipitation will occur, in the form of rain, hail, snow or sleet.
4. Some of the water that falls is intercepted by the leaves on trees.
5. Some of the water will be stored on the surface (particularly if it is snow), will infiltrate into the soil or will flow over the land as surface run-off.
6. Some of the water that travelled as infiltration will move horizontally through the soil as throughflow.
7. Some of the water will move down through permeable rocks in a process called percolation.
8. Some of the water is stored as groundwater in porous rocks.

You need to know how to draw this diagram.

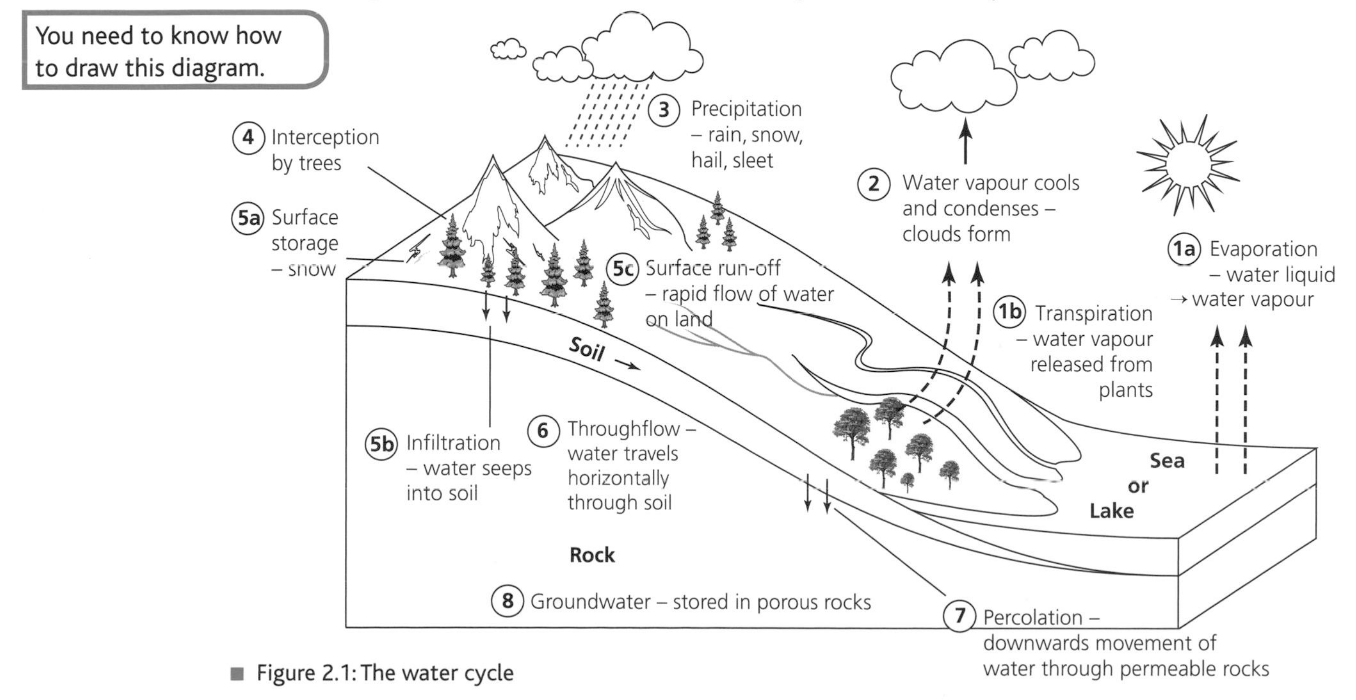

■ Figure 2.1: The water cycle

Revision tip

You could try making hand gestures for each of the transfers, flows and processes in the water cycle. For example, evaporation could be a flat hand rising, for condensation you could close your fist to resemble a water droplet, and for interception your fist could tap your head. You could make up the hand gestures with a friend and then test each other.

2.2 Rainfall types

There are three main types of rainfall that are experienced in the UK: relief, convectional and frontal.

Relief rainfall

1. Evaporation causes warm, moist air over the sea.
2. As the air meets a hill, it is forced to rise.
3. As it rises, the air cools and then condenses at the dew point.
4. Clouds form and rain falls.
5. The air sinks over the other side of the hill. No rain falls here in the rainshadow.

You need to know how to draw this diagram.

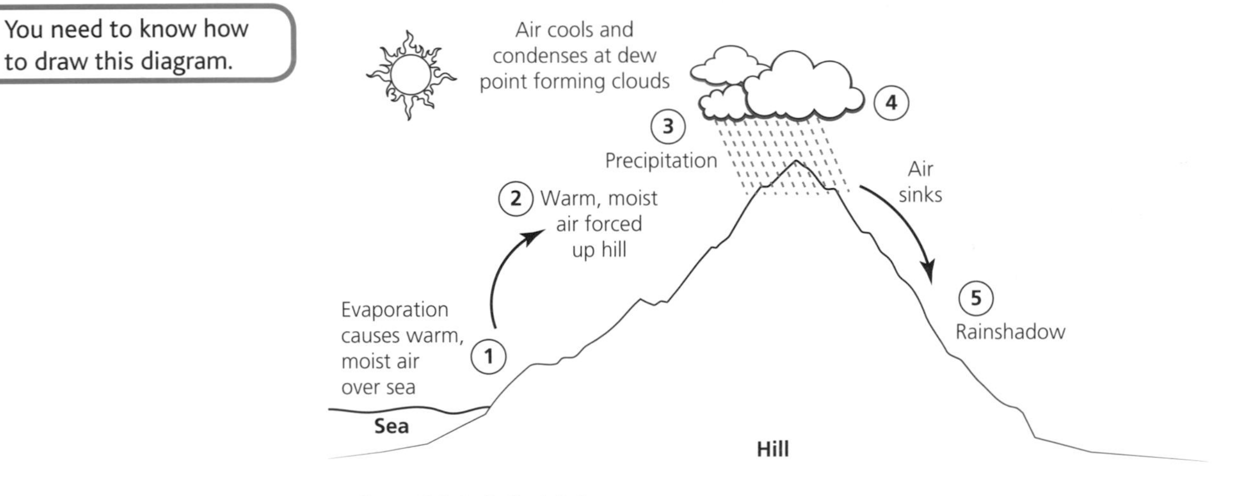

■ Figure 2.2: Relief rainfall

Relief is the term used in geography to describe the shape of the land.

Relief rainfall occurs in hilly or mountainous places, such as Wales, Scotland, the Alps and the Rockies.

Places at the foot of hills or mountains which do not face the prevailing wind are in the rainshadow and do not get very much rainfall.

Convectional rainfall

1. Hot sun heats any water on the ground.
2. Water from the ground is evaporated.
3. Water vapour rises, cools and condenses at the dew point.
4. Clouds form and rain falls.

You need to know how to draw this diagram.

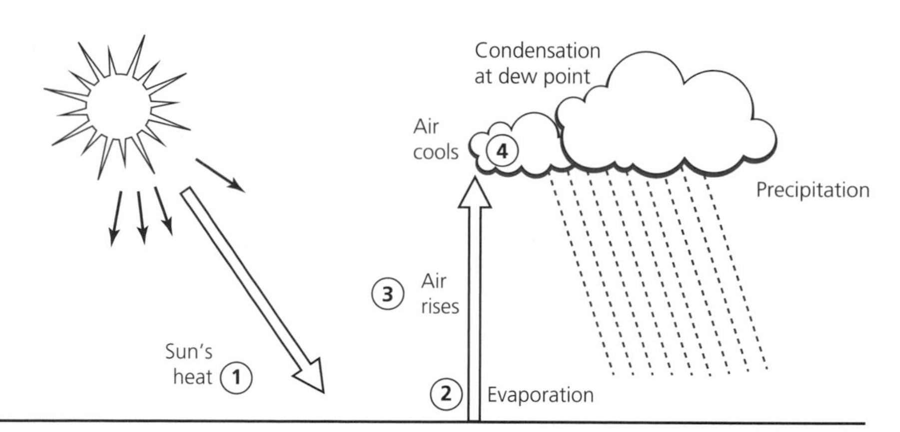

■ Figure 2.3: Convectional rainfall

Convectional rainfall occurs in places that have strong sunshine and are relatively near a sea, lake or ocean.

Britain can experience convectional rain in the summer when it is very hot.

Tropical rainforests get convectional rain every day. The sun in the morning heats the puddles on the ground from the previous day's rain, then, by midday, it rains again.

Frontal rainfall

1. A warm air mass meets a cold air mass. The boundary where they meet is called a front.
2. Cold air is heavier, so it undercuts the warm air.
3. The warm air rises, cools and condenses at the dew point.
4. Clouds form and it rains.

You need to know how to draw this diagram.

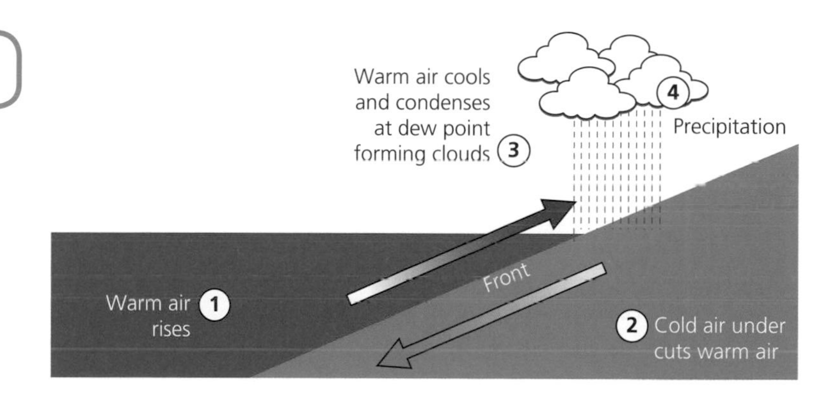

■ Figure 2.4: Frontal rainfall

Frontal rain occurs in places where air masses from tropical areas and polar areas meet.

Britain receives much frontal rain.

When hot air and cold air meet, air pressure is low, as air is rising. This weather system is called a depression and brings very changeable weather.

Revision tip

You could make revision cards for each of the rainfall types, ensuring that you add a diagram to each card. You could also use hand gestures to remember why the warm air rises for each of the rainfall types. For relief rainfall draw a hill with your finger in the air, for frontal rainfall put your arm diagonally in the air to represent a front, for convectional rainfall flash your hand in the air to represent hot sun. Test each other on these hand gestures!

2.3 Factors affecting temperature

There are a number of factors that affect temperature.

Latitude

The temperature rises as you get closer to the Equator, and falls as you get closer to the poles. This is because the Sun's rays have further to travel to get to the poles. For this reason, the south of Britain is warmer than the north.

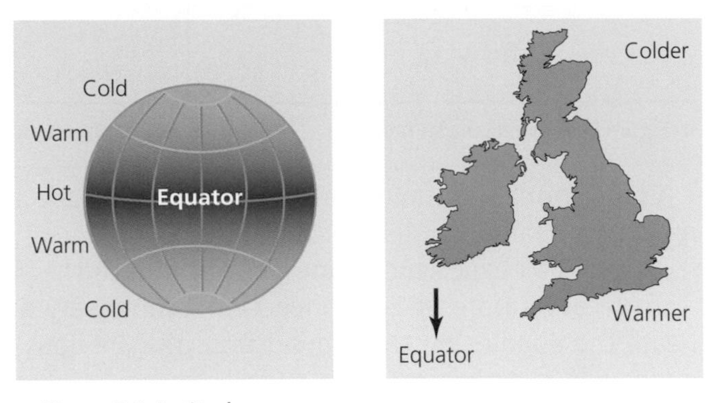

■ Figure 2.5: Latitude

Altitude

The height of the land (the altitude) affects the temperature. The temperature falls by approximately 1°C for every 150 m you ascend.

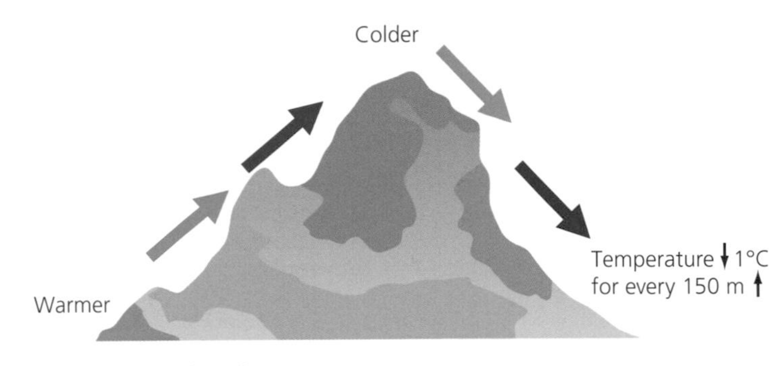

■ Figure 2.6: Altitude

Distance from the sea

During the summer, the further inland you go, the warmer it gets; during the winter, the opposite is true.

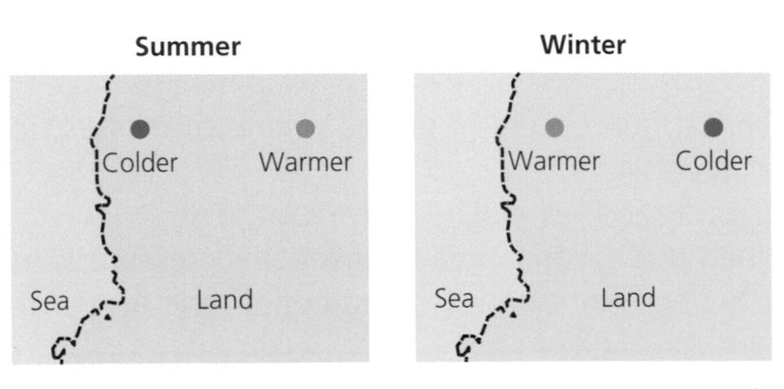

■ Figure 2.7: Distance from the sea

The sea is very deep, so it takes a long time to heat up, but once it is warm it takes a long time to cool down. (Think of it being like a lasagne that takes a long time to cook, and then a long time to cool down.) The land is quick to heat up, but cools down quickly too. (It is like cheese on toast that just gets grilled on top and is quick to cook but cools down very quickly once you leave it on the table!)

Ocean currents

In the UK, the North Atlantic Drift, or Gulf Stream, means that the west side of the country is warmer than the east side. The Gulf Stream is an ocean current which affects Western Europe, increasing the temperature by several degrees in winter. It makes Britain warmer in winter than other places at the same latitude of 50–60 degrees north. It reduces frosts and keeps waterways and ports in Western Scotland ice free.

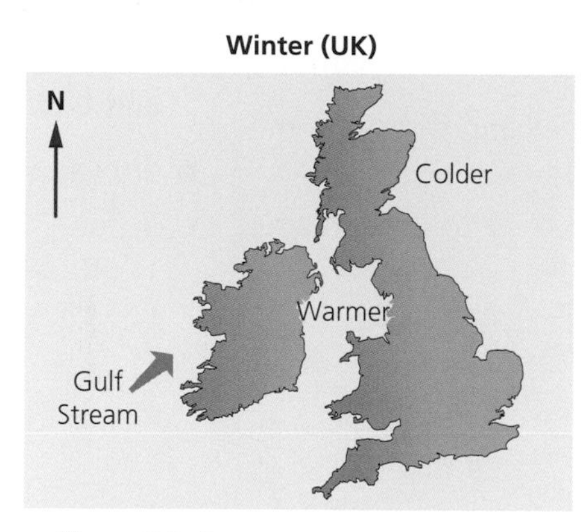

■ Figure 2.8: Ocean currents

Prevailing wind direction

In the UK, the wind blows from the south-west for 80 per cent of the time. This is a warm wind. When the wind blows from the south it is a warm/hot wind from Northern Africa. When the wind blows from the north it is a cold wind.

Jet streams affect our weather conditions. A jet stream is a strong flowing ribbon of air high up in the atmosphere; it can blow at 160 km per hour. Depending on its position it can bring warmer, colder, wetter or windier weather. It can push depressions (weather systems that bring wind and rain) towards the British Isles. The storms of 2007, 2012 and the winter of 2013–14 were caused by a jet stream that normally tracks north of the British Isles shifting south and pushing depressions over the British Isles.

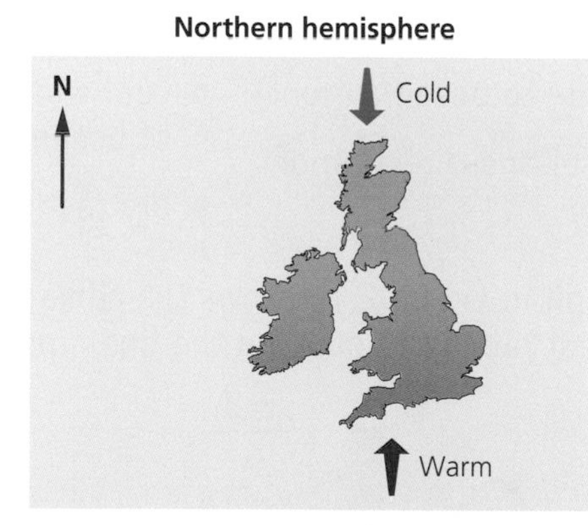

■ Figure 2.9: Prevailing wind direction

Revision tip

The first letter of each of these factors that affect temperature spells a made-up word **LADOP**. You may find this an easy word to remember or you could change it to **DOPLA**. Alternatively, you could make up a mnemonic such as '**l**adies **a**lways **d**ance **o**n **p**lates'!

2.4 A humid temperate climate (Britain)

Britain has a humid temperate climate. This means that it usually has warm summers and mild winters, and rainfall throughout the year. The climate varies from one region to another.

North west
- mild summers (due to latitude)
- mild winters (due to ocean current)
- wet (due to relief and direction of prevailing wind)

North east
- mild summers (due to latitude)
- very cold winters (due to latitude and lack of ocean current)
- dry (as in rainshadow)

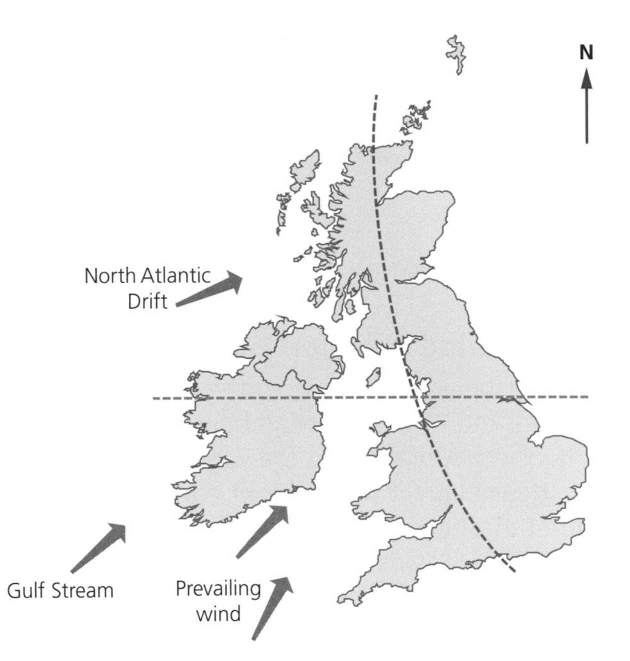

■ Figure 2.10: Britain's climate

South west
- warm summers (due to latitude)
- mild winters (due to ocean current)
- wet (due to relief and direction of prevailing wind)

South east
- warm summers (due to latitude)
- cold winters (due to lack of effect of ocean current)
- dry (as in rainshadow)

The climate graph in Figure 2.11 shows the climate for a typical UK location. The blocks represent rainfall in mm and the line represents temperature in °C.

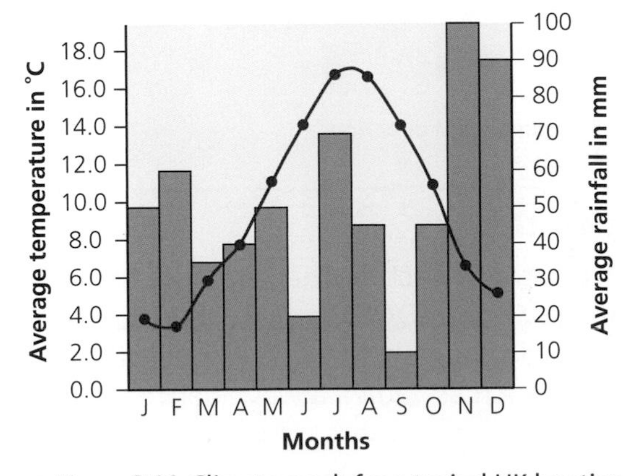

■ Figure 2.11: Climate graph for a typical UK location

The map in Figure 2.12 shows the average rainfall for Great Britain. This is a choropleth map. The areas coloured darkest have the most rainfall and those shaded a light colour have least rainfall.

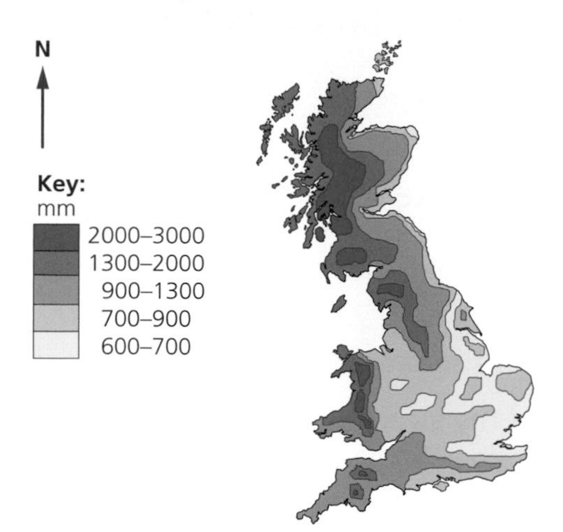

■ Figure 2.12: Map of Great Britain showing average rainfall per annum

2.5 A humid tropical climate

This is included for comparison but it is not part of the syllabus.

A humid tropical climate is experienced in equatorial regions of the world: West Africa, South East Asia, Northern Australia and South America (for example, the Amazonian rainforest in Brazil).

- Average annual rainfall is over 2000 mm.
- The range of temperature over the year is 1°C. The temperature is not seasonal.

The humid tropical climate is caused by:

- The equatorial location. The Sun is therefore overhead for most of the year. The rays are concentrated on a small area causing high temperatures.
- Convectional rainfall. The Sun in the morning evaporates the water. Storm clouds form and heavy rainstorms occur in the afternoon.

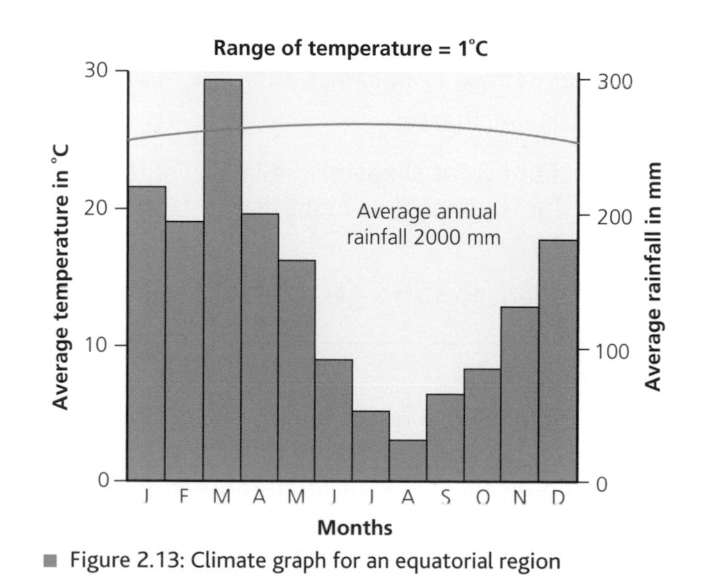

■ Figure 2.13: Climate graph for an equatorial region

Exam-style questions

Try these questions. Answers are near the back of the book.

2.1 Describe how a humid tropical climate differs from a humid temperate climate. (3)

2.2 Explain these differences. (3)

2.6 Microclimates

A microclimate is the local climate of a small area. A number of factors affect microclimate:

- The physical features of the area, such as hills, lakes, valleys.
- The aspect of the area. This is the direction that a slope or wall faces; a south-facing slope is the warmest.
- The wind direction. In the UK, a northerly wind will be colder than wind from any other direction.
- Proximity to buildings. Buildings release heat and can provide shelter, thus increasing temperatures.
- The surface. Dark surfaces absorb heat.
- The distance from the sea. In winter, places near the sea are warmer than those further from the sea.

The microclimates in urban areas differ from those in rural areas.
An urban microclimate:

- is 1°C warmer than rural areas during the day. This is partly because man-made heat is released from power stations, houses, cars, etc.
- is 4°C warmer than rural areas at night. Tarmac absorbs heat during the day and releases it at night
- experiences less wind than rural areas. Tall buildings act as wind breaks, but funnelling between buildings can cause gusts
- has more convectional rainfall than rural areas because it is warmer than rural areas
- has less snow.

A rural microclimate:

- is affected by the shape of the land. South-facing slopes are warmer and valley floors are cold at night due to cold air sinking (sometimes causing frosts)
- often experiences stronger winds because there is less shelter.

Revision tip

You could make a mind map to revise the whole weather and climate topic. Write 'weather and climate' in the centre and then have sticks coming off for the water cycle, rainfall types, LADOP, Britain's climate and microclimates. Each of these categories could then have sticks coming off them explaining all of the important points. Remember to use memory pictures rather than words to represent the facts. Use lots of colours but try to use colour to good effect.

Exam-style questions

Try these questions. Answers are near the back of the book.

2.3 The local atmospheric conditions in a small area, such as the grounds of a school, are called:
(a) weather
(b) precipitation
(c) microclimate. (1)

2.4 Name three factors that may be very important in influencing the local climate (for example, within the school grounds). (3)

2.5 Why might the local climate vary during the course of a bright, sunny day? (4)

★ Make sure you know

- ★ The process of the water cycle
- ★ The different types of rainfall
- ★ The different factors affecting temperature
- ★ The reasons why temperature and rainfall vary across Britain
- ★ The factors affecting microclimates

Test yourself

Before moving on to the next chapter, make sure you can answer the following questions. Answers are near the back of the book.

1 **(a)** Name two factors that affect temperature.

____

____

(b) What is the difference between weather and climate?

____

____

2 **(a)** What do geographers call the process of water sinking into soil?

____

____

(b) What do geographers call the process of water travelling over the top of soil?

____

____

3 Write out each of these sentences using the correct word or words to finish the sentence:

(a) Rain, hail, snow and sleet are all forms of

infiltration **precipitation** **rainfall** **weather system**

(b) The prevailing wind direction for the UK is from the

south-east **north-west** **south-west** **north-east**

(c) The Gulf Stream is

an ocean current **a wind** **a river** **an island**

(d) The Gulf Stream and North Atlantic Drift affect the south-west of the UK

in summer **all the year round** **in spring** **in winter**

(e) Moist air forced to rise over upland areas causes

relief rainfall **frontal rainfall** **convectional rainfall**

4 Write the definitions of these words and phrases you need to know, and then ask someone to check them. (Words and phrases in italics are useful even though you are not required to know them.)

air mass

____

____

dew point

____

____

altitude

____

____

drought

____

____

anticyclone

____

____

Equator

____

____

aspect

____

____

evaporation

____

____

atmosphere

____

____

fog

____

____

climate

____

____

front

____

____

condensation

____

____

frontal rainfall

____

____

convectional rainfall

____

____

Gulf Stream

____

____

depression

____

____

hemisphere

____

____

humidity

precipitation

infiltration

prevailing wind

interception

rainshadow

irrigation

relief rainfall

isotherm

surface run-off

latitude

throughflow

microclimate

transpiration

North Atlantic Drift

weather

percolation

Rivers and coasts

In this chapter, you will be looking at the processes of weathering and erosion and how these create various landforms.

- **Weathering** is the breaking down of rocks by weather, plants and animals.
- **Erosion** is the wearing away and removal of rocks by rivers, sea, ice and wind.

3.1 Rock types

Weathering and erosion work at different speeds on different types of rock.

Igneous rock

This is formed from volcanic rock. If the magma cools underground, granite is formed. If it reaches the Earth's surface, it is called lava, which then forms basalt when it cools.

Sedimentary rock

This is formed when rivers transport particles of rock and remains of plants and animals to the sea. These then sink to the sea bed and, over millions of years, compress to form new rock.

Metamorphic rock

This is formed from sedimentary or igneous rock when it is exposed to extreme pressure or heat during the Earth's movements, for example, chalk and limestone turn to marble; clay turns to slate.

Revision tip

Try drawing a flow chart to show the rock cycle. You could show igneous rock coming out of a volcano, then being eroded by a river and being taken to the sea to form sedimentary rock. This could then be dragged down by subduction at a destructive plate boundary and then be compressed underground to form metamorphic rock.

3.2 Types of weathering

There are three main types of weathering: physical, chemical and biological. And there are two sorts of physical weathering: freeze-thaw and exfoliation.

Physical weathering: freeze-thaw weathering

- This process starts when water seeps into cracks in the rock.
- At night the temperatures fall below 0 °C, the water freezes and, as ice, expands.
- This forces the cracks open.

- The process happens again and again, and breaks up the rock.
- The loose rock is called scree.

Freeze-thaw is common in mountainous areas. Igneous rocks (granite) and metamorphic rocks (marble) from uplands are prone to this type of weathering.

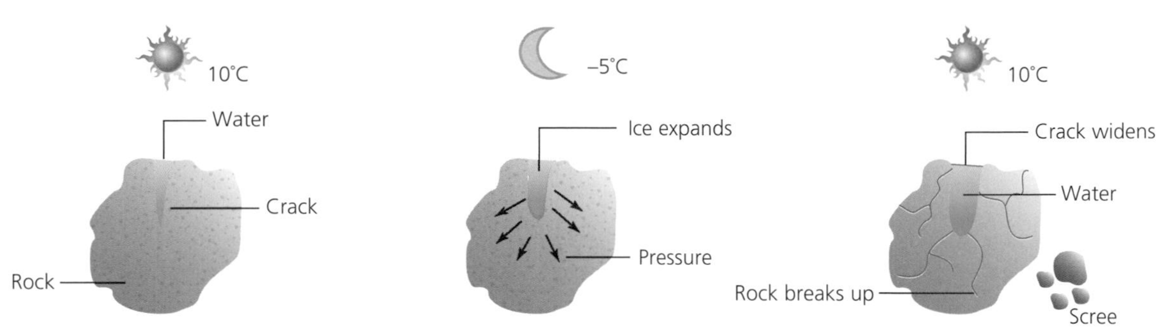

■ Figure 3.1: Freeze-thaw weathering

Physical weathering: exfoliation

- This process happens when rocks are repeatedly subjected to heat and cold.
- Heat from the Sun makes the outer layer expand.
- The cold at night makes the outer layer contract.
- The outer layer of the rocks then peels off.
- The loose rock is called scree.
- This type of weathering is common in desert areas, which are hot in the day and cool at night.

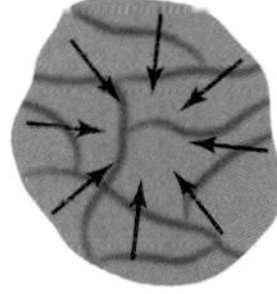

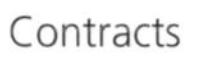

■ Figure 3.2: Exfoliation

Biological weathering

- This process is caused by plants and animals.
- Burrowing animals break up the rocks.
- Plant seeds fall into cracks and germinate, breaking up the rocks.
- Tree roots grow into cracks in the rocks and then exert pressure on the cracks as they grow, causing them to widen.

Chemical weathering

- This process is caused by rain, which contains carbonic acid.
- The acid in the rain attacks the rock, causing it to crumble.

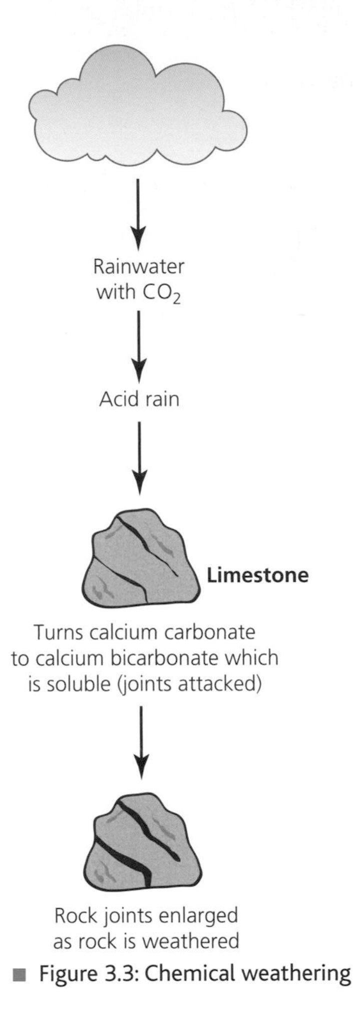

■ Figure 3.3: Chemical weathering

Sedimentary rocks, such as limestone and chalk, are particularly vulnerable to this type of weathering. As the carbonic acid falls on limestone, it turns it into calcium bicarbonate, which is soluble in water.

Limestone gravestones are commonly attacked and limestone pavements are also vulnerable as the acid water can seep into the grykes (deep cracks) and attack a large surface area.

Revision tip

Try drawing a flow chart for each of the weathering types. For instance, for freeze-thaw weathering you could use:

- Water into crack in rock in mountain
- Water freezes
- Ice has greater volume than water so puts pressure on rock
- Rock begins to break up and form scree

3.3 Features of the river basin

The area of land drained by a river and its tributaries is called the drainage or river basin.

- The watershed is the edge of the river basin.
- The source is where the river starts.
- Tributaries are small streams or rivers than run into the main river channel.
- The confluence is the point where the tributary meets up with the main channel.
- A meander is a bend in the river.
- An ox-bow lake is where a meander has been cut off and a small lake has formed.
- The flood plain is the flat land in the lower course of the river, which is prone to flooding.
- The estuary is a wide area near the mouth of the river where the sea (salt) water mixes with fresh water, forming brackish water.
- The mouth is the point where the river reaches the sea.
- At this point, the river drops much of the load it has been carrying, which sometimes forms a delta.

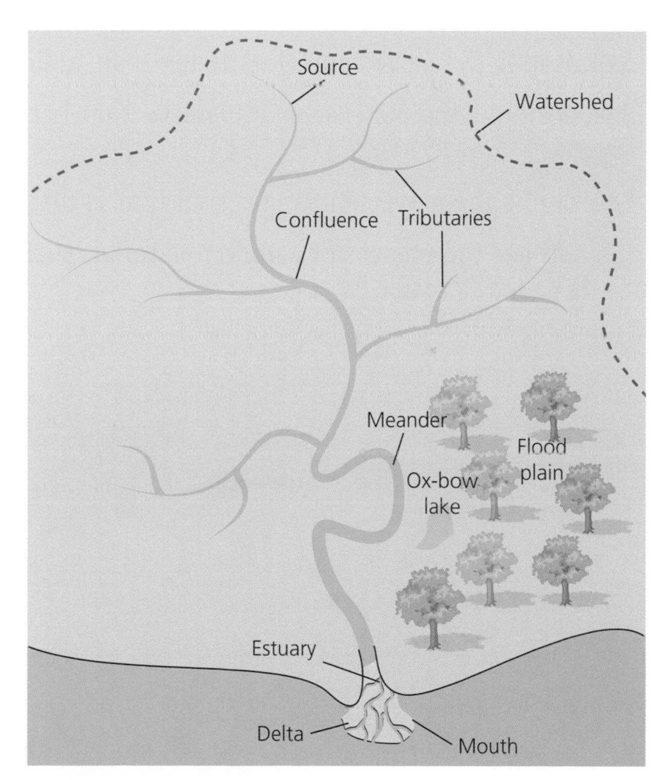

■ Figure 3.4: River basin

Revision tip

You could make flash cards to revise the features of a river basin. On one side write the name of the feature and on the other side draw a memory picture to represent the word.

3.4 River processes

Three processes occur as a river flows from its source to its mouth. These are erosion, transportation and deposition. There are four types of erosion and four types of transportation.

Erosion

As the river moves through the river basin, it alters the landscape due to the wearing away and removal of land caused by the following processes:

- Attrition occurs when particles of load collide and knock pieces off each other.
- Abrasion occurs when smaller material rubs against the bed and banks of the river.
- Corrosion occurs when acid in the water dissolves particles of rocks from the bed and banks of the river.
- Hydraulic action is the sheer force of the water and air forcing itself into the soil and moving away parts of the bed and banks of the river.

Transportation

Once the material (known as load) has been eroded, it is then carried along the river by the following processes:

1. Traction is the rolling of stones along the river bed.
2. Saltation is the movement of particles 'leap-frogging' along the river bed.
3. Suspension is the movement of material that is carried within the water flow.
4. Solution is the movement of material that is dissolved in the water.

(Sometimes the process of carrying material, such as twigs, on top of the water is called flotation.)

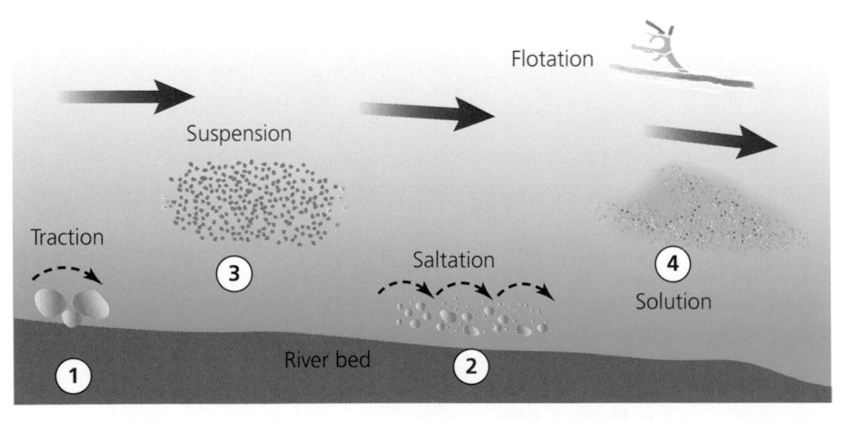

Figure 3.5: Transportation

Deposition

When the river slows down, the load is 'dumped'. This is known as deposition. Large boulders are deposited first and fine sediment last.

Revision tip

The types of transportation and erosion can really easily be represented by hand gestures. You could decide with your friends what gesture should represent each process. For example, attrition could be your two fists tapping each other (each fist represents a piece of load), abrasion could be one fist tapping a flat hand (the fist represents a piece of load and the flat hand represents the bank). Once you have decided on all the gestures, you could test each other.

3.5 Features of the upper course

V-shaped valleys and waterfalls are the main features of the upper course of a river.

V-shaped valley

This is a feature of erosion. It occurs when the river erodes downwards into the land by abrasion and hydraulic action. This is called vertical erosion. The valley sides are then shaped by the weather, plants and animals (weathering).

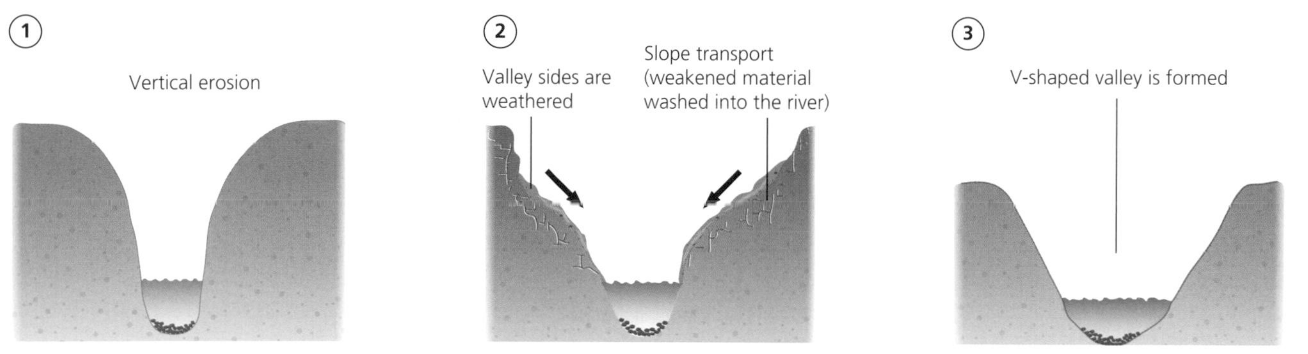

■ Figure 3.6: V-shaped valley formation

Waterfall

This is a feature of erosion. It occurs when a river flowing over hard rock meets a band of softer, less resistant rock.

- Hydraulic action and abrasion erode the softer rock forming a 'step' in the river bed.
- The softer rock is undercut and the hard rock is left as an overhang. A plunge pool is formed at the base of the waterfall; this plunge pool is deepened by hydraulic action, corrosion and abrasion as the pebbles erode its base.
- The overhang eventually collapses and in this way the waterfall retreats towards the source of the river.
- As the erosion continues, a gorge is formed which is a steep-sided valley.

You need to know how to draw this diagram.

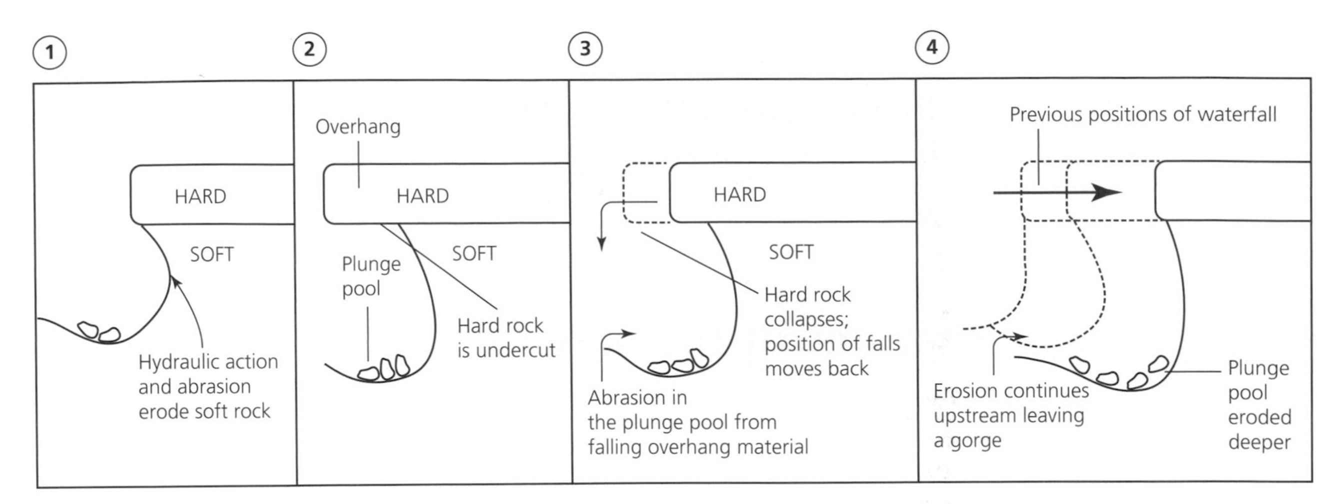

■ Figure 3.7: Waterfall formation

3.6 Features of the lower course

Meanders, ox-bow lakes, deltas and flood plains are the main features of the lower course of a river.

Meanders

This is a feature of erosion and deposition. The river is dynamic – it is constantly changing its shape and therefore has a lot of meanders (bends) in it. These meanders are formed by lateral (sideways) erosion.

Ⓐ Outside of a meander

- river cliff
- fast velocity (*= fastest flow (thalweg))
- erosion (hydraulic action and abrasion)
- deeper water

Ⓑ Inside of a meander

- river beach / slip-off slope
- slow velocity
- deposition
- shallow water

You need to know how to draw this diagram.

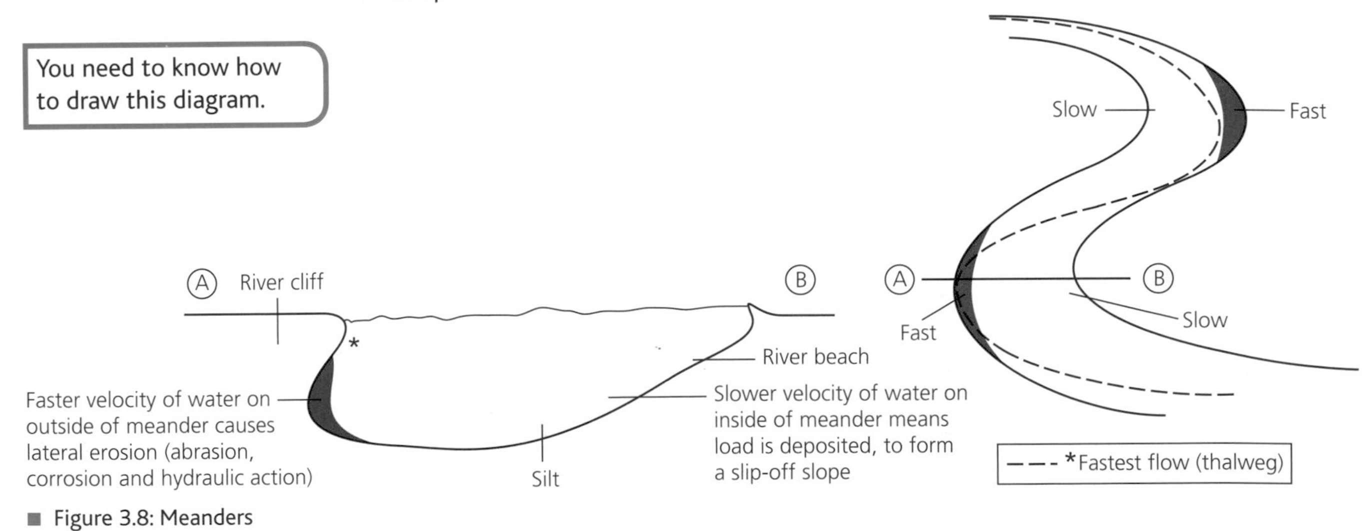

■ Figure 3.8: Meanders

Ox-bow lake

This is a feature of erosion and deposition. It occurs where the horseshoe-shaped meander becomes tighter, until the ends become very close together and join to form a separate lake.

① The outsides of two meanders are eroded by hydraulic action and abrasion.

② The river becomes more sinuous (has more curves and turns). This results in a narrow neck of land remaining between the two river cliffs.

③ Eventually, perhaps during a flood, the narrow neck of land is eroded away and the water takes the more direct straight route downstream.

④ Deposition occurs and eventually the old meander loop is separated from the river and forms an ox-bow lake.

Evaporation will usually cause the lake to become dry eventually.

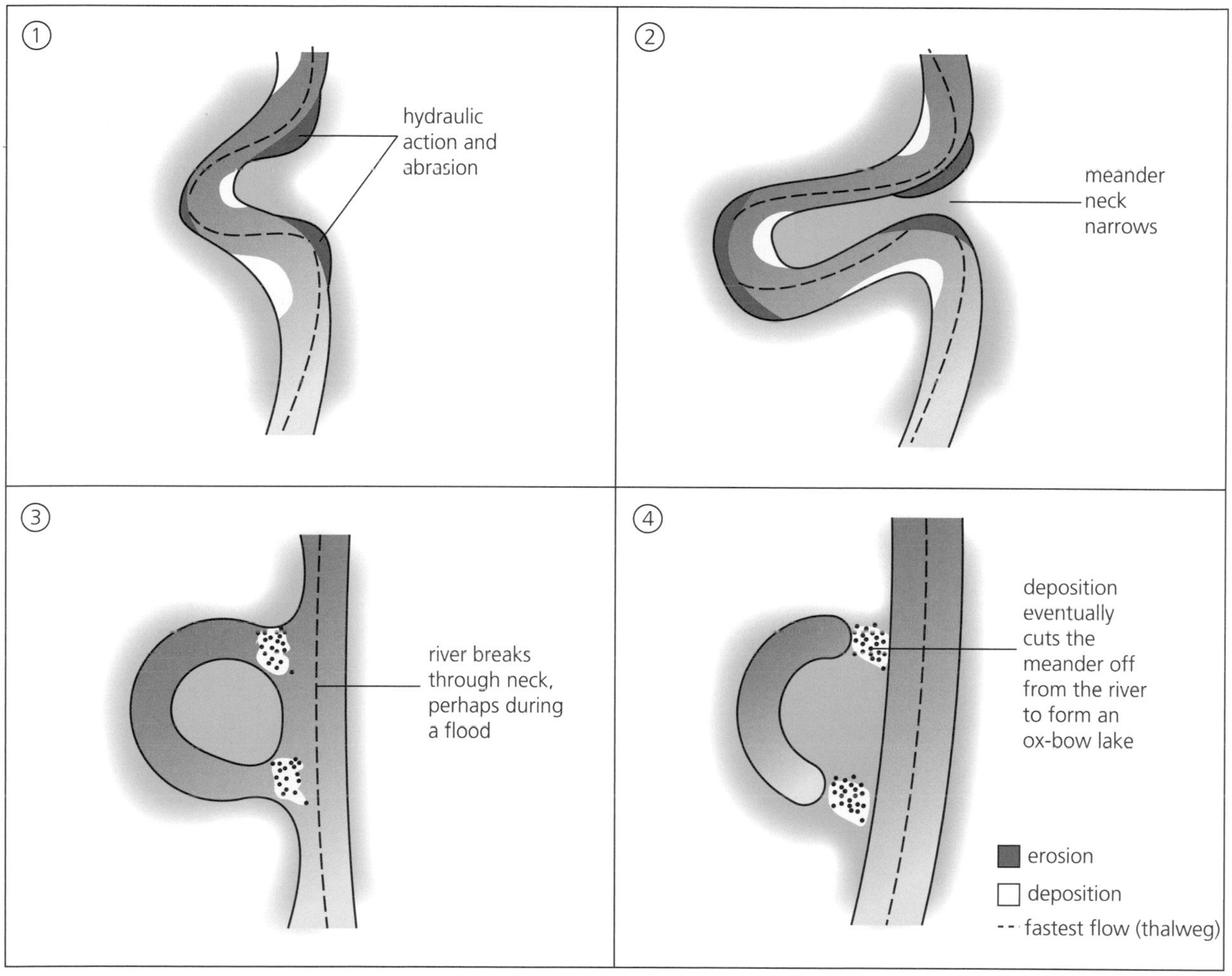

■ Figure 3.9: Ox-bow lake formation

Deltas

This is a feature of deposition. Examples are the Nile Delta and the Mississippi Delta.

- As large rivers approach the sea, they carry a large amount of load (material) in suspension.
- The speed (velocity) of the river is reduced as it reaches the more powerful sea, so it has less energy and deposits its load to form new land.
- The coarser material is deposited first (in topset beds), the medium sized silt (in foreset beds) and then the finer material (in bottomset beds).

- With time, more and more sand and silt is deposited.
- The river divides into channels called distributaries which flow round the deposits of new land.

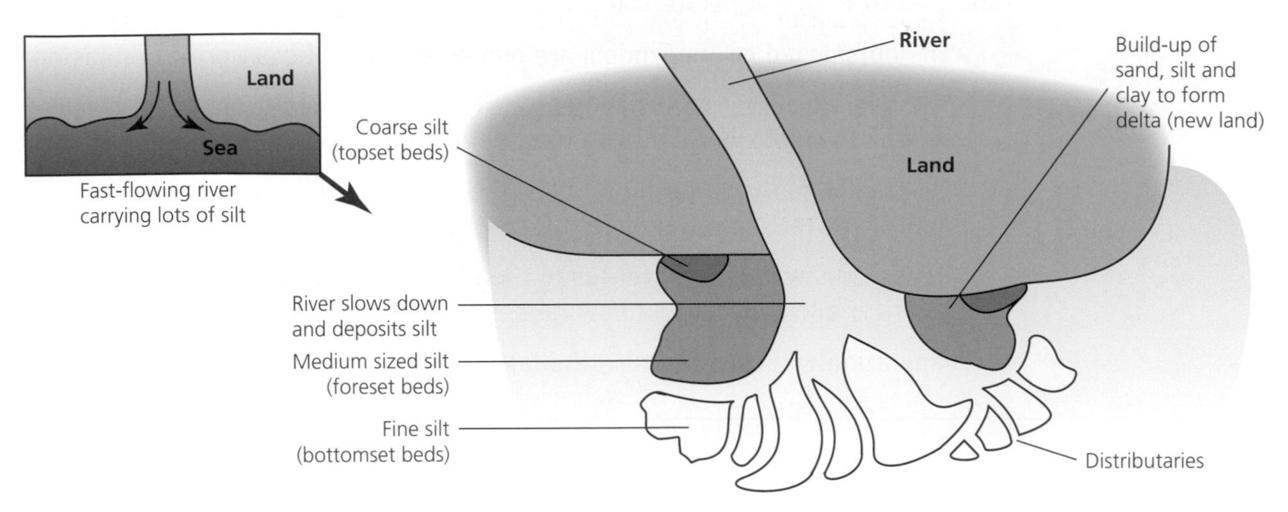

■ Figure 3.10: River delta formation

Flood plain

This is a feature of deposition. It occurs when a river floods and deposits its load.

- As the water spills out of its channel, friction increases, the water velocity decreases and deposition occurs.
- The larger pieces of load are deposited first, often forming natural levees near the channel.
- Finer sediment is transported further away from the channel.
- The flat land onto which flood water flows is known as the flood plain.

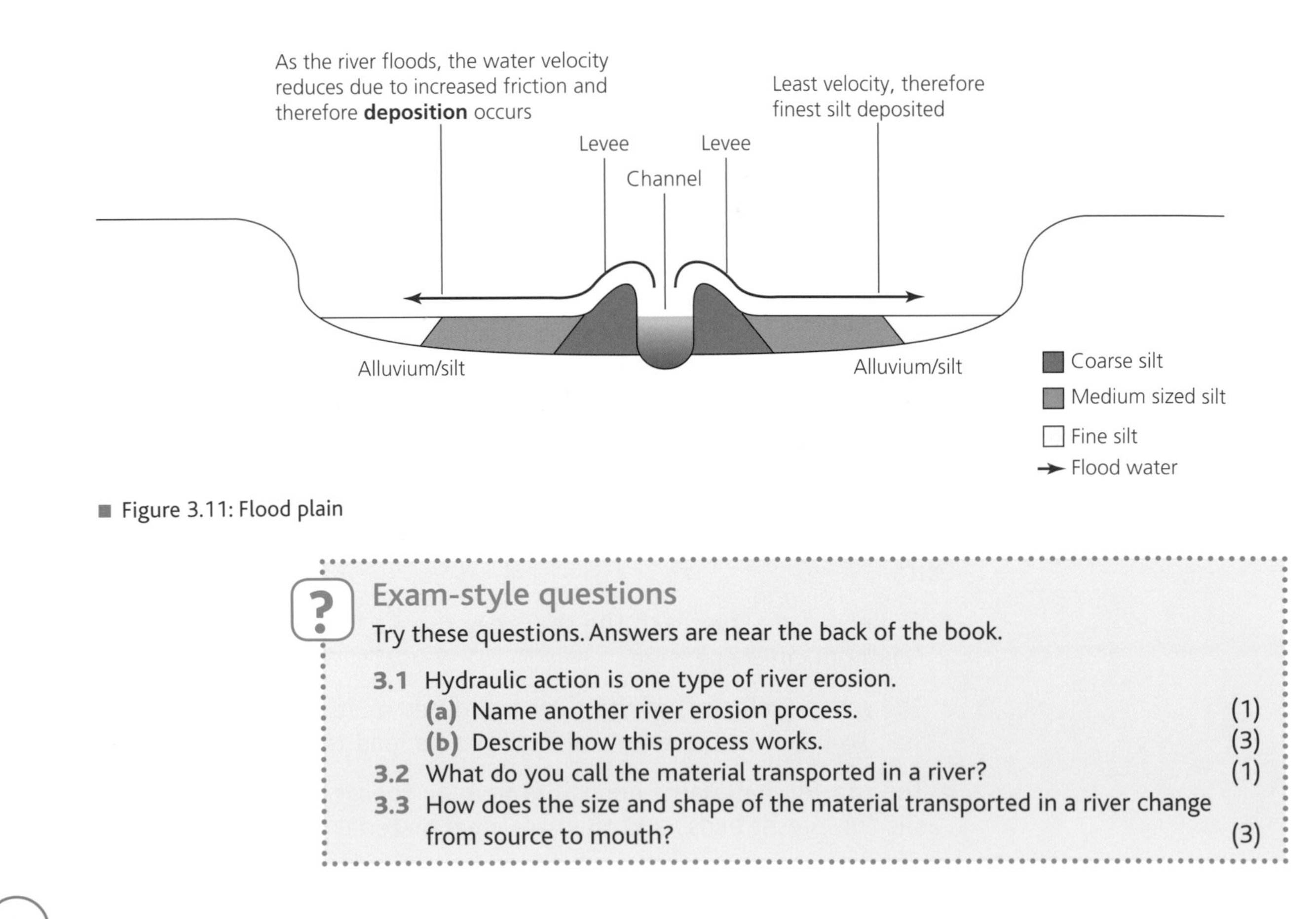

■ Figure 3.11: Flood plain

Exam-style questions

Try these questions. Answers are near the back of the book.

3.1 Hydraulic action is one type of river erosion.
  (a) Name another river erosion process. (1)
  (b) Describe how this process works. (3)

3.2 What do you call the material transported in a river? (1)

3.3 How does the size and shape of the material transported in a river change from source to mouth? (3)

3.7 Coastal erosion

There are four types of coastal erosion.

- Hydraulic action is the force of the waves against the cliffs. The water traps air in cracks and caves. The air is compressed, forcing the rock to weaken and eventually break.
- Corrosion is caused by the acid in the seawater spray dissolving the rock.
- Attrition is caused by pebbles hitting each other in the waves. This makes the pebbles smaller and rounder and eventually they become sand.
- Abrasion is the effect of waves throwing pebbles at the cliffs. This erodes the cliffs at their base.

The features of erosion are:

- headlands and bays
- caves, arches, stacks and stumps.

Headlands and bays

These are features of erosion and deposition.

①

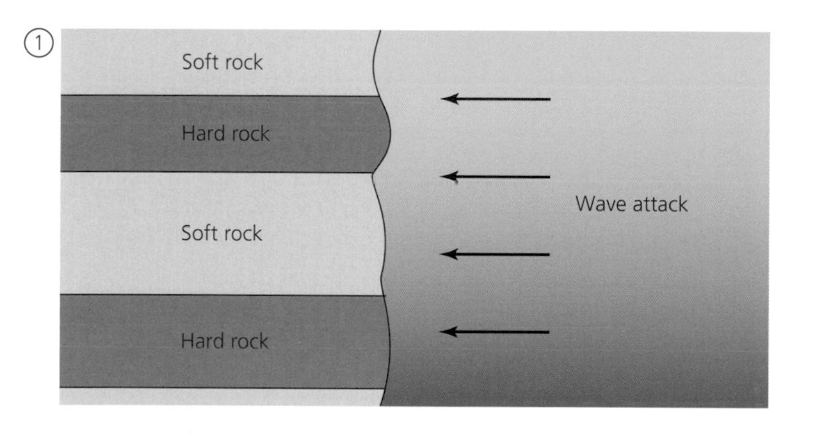

②

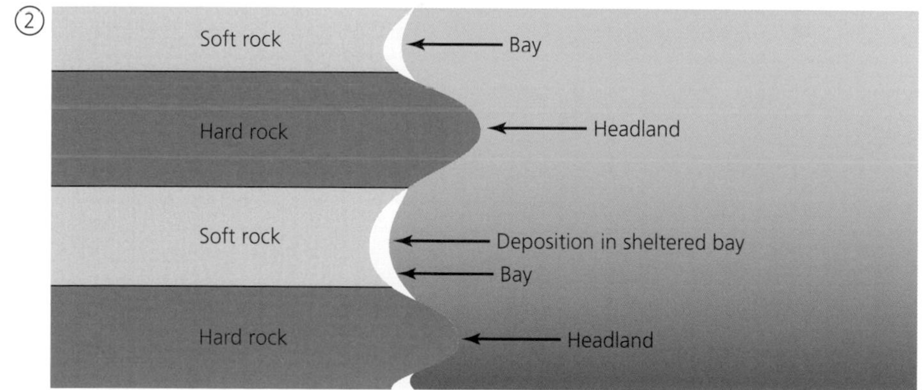

■ Figure 3.12: Headland and bay formation on discordant coastline

① Waves attack a discordant coastline of alternating hard and soft rock. The soft rock erodes at a faster rate than the hard rock.

② Headlands are created from the hard rock and bays are eroded from the soft rock between them. Deposition also occurs in sheltered bays.

Caves, arches, stacks and stumps

These are features of erosion. See Figure 3.13.

1. Waves attack a fault in the rock by hydraulic action and abrasion.
2. The fault is enlarged to form a cave. A blowhole may appear on the headland due to upward erosion by waves on the roof of the cave.
3. Hydraulic action and abrasion widen and deepen the cave and eventually cut through the headland to form an arch.
4. Undercutting, weathering and gravity lead to collapse, leaving a stack.
5. Weathering and erosion turn the stack into a stump.

As the cliff retreats through erosion by waves, a platform of rock extending into the sea may be left. This is called a wave cut platform.

Exam-style question

Try this question. The answer is at the back of the book.

3.4 Describe the processes involved in the erosion of a headland. (5)

You need to know how to draw this diagram.

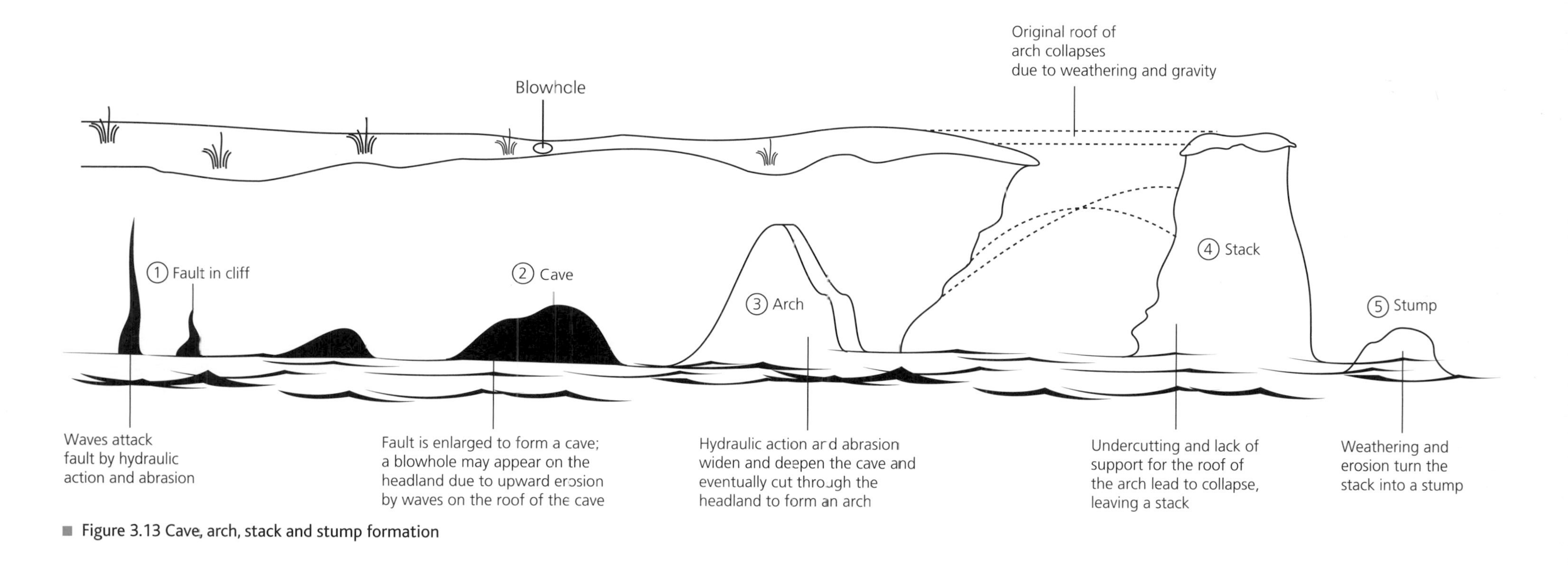

Original roof of
arch collapses
due to weathering and gravity
Blowhole
1 Fault in cliff
2 Cave
3 Arch
4 Stack
5 Stump
Waves attack fault by hydraulic action and abrasion
Fault is enlarged to form a cave; a blowhole may appear on the headland due to upward erosion by waves on the roof of the cave
Hydraulic action and abrasion widen and deepen the cave and eventually cut through the headland to form an arch
Undercutting and lack of support for the roof of the arch lead to collapse, leaving a stack
Weathering and erosion turn the stack into a stump

■ Figure 3.13 Cave, arch, stack and stump formation

3.8 Coastal transportation

Longshore drift is the movement of sediment along the beach by waves.

- The swash moves up the beach at the angle determined by the direction of the prevailing wind.
- After the wave has broken, the backwash returns to the sea at a 90° angle.
- In this way, sand and pebbles are moved along the beach.
- The sand and pebbles will build up against a groyne.

You need to know how to draw this diagram.

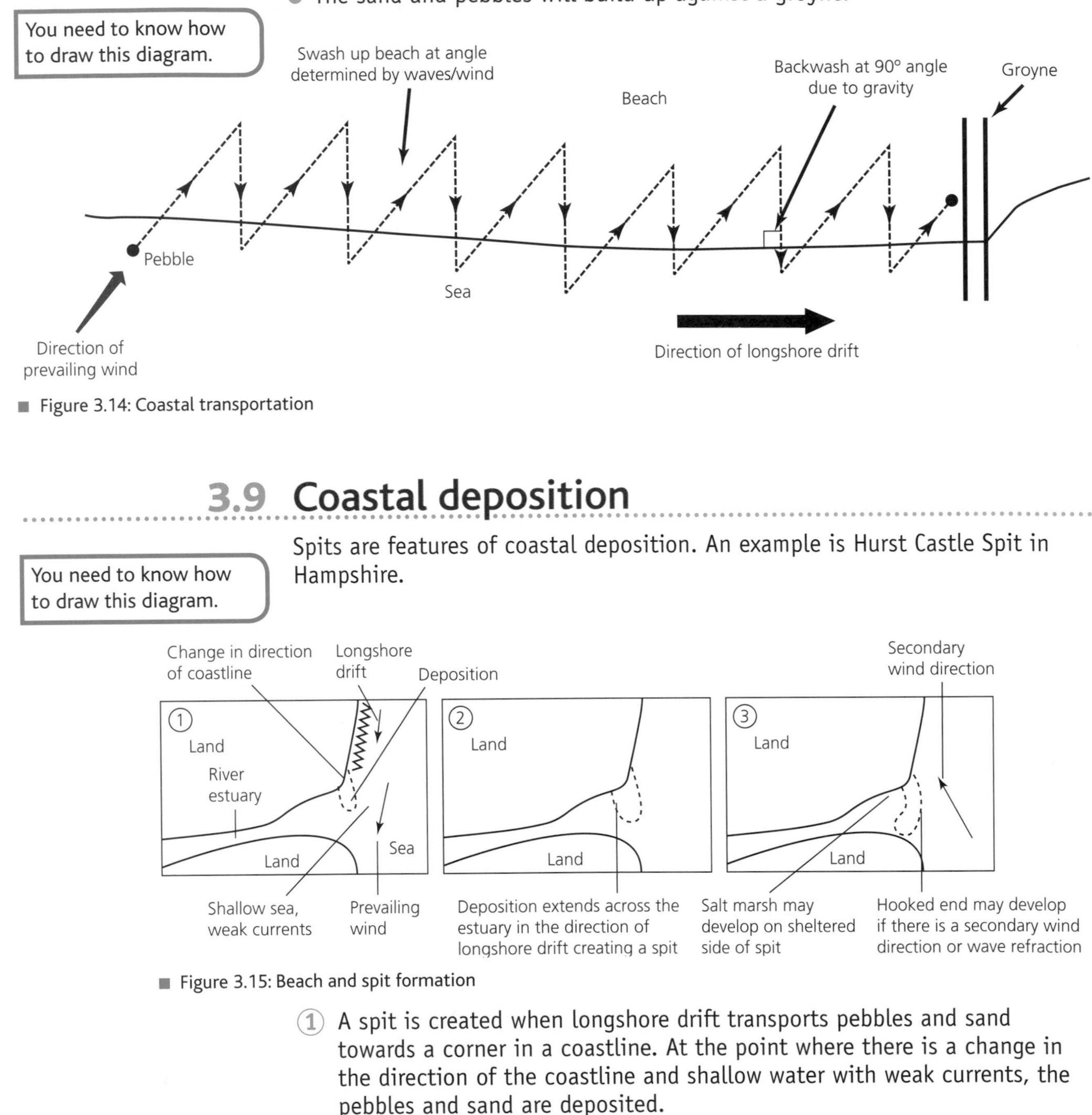

■ Figure 3.14: Coastal transportation

3.9 Coastal deposition

You need to know how to draw this diagram.

Spits are features of coastal deposition. An example is Hurst Castle Spit in Hampshire.

■ Figure 3.15: Beach and spit formation

① A spit is created when longshore drift transports pebbles and sand towards a corner in a coastline. At the point where there is a change in the direction of the coastline and shallow water with weak currents, the pebbles and sand are deposited.

② Over time the deposition will extend into the sea in the direction of the longshore drift.

③ A salt marsh will then develop on the sheltered side of the spit. The spit may develop a hooked end if there is a secondary wind direction or wave refraction.

Revision tip

Try using play dough to make models of all of the features (spits, stacks and stumps, waterfalls, meanders and flood plains). Don't just make the model but also show how the feature changes over time and give a talk about what processes are occurring as you manipulate the dough. Remember to use as much geographical terminology as possible.

Exam-style question

Try this question. The answer is at the back of the book.

3.5 Describe and draw the processes involved in the formation of a spit. (5)

3.10 Causes of flooding – river and coastal

The causes of flooding can be broken down into: climatic, physical and human.

Climatic

- Heavy rainfall over a short or long period of time
- Ground that is saturated as a result of previous rainfall
- Melting snow and glaciers
- Soil baked hard by the Sun therefore acting as if impermeable
- Hurricanes causing wave surges

Physical

- Narrow steep-sided valleys causing surface run-off to reach rivers rapidly after a storm
- Impermeable rock causing rapid surface run-off
- Low-lying coastal areas
- Tsunamis

Human

- Urbanisation can lead to an increase in tarmac and drains, which causes rapid surface run-off.
- The diversion of a river or the narrowing of a channel can lead to flooding.
- Deforestation in a river basin leads to less water being taken up by roots.
- Deforestation and the loss of tree roots mean that the soil is loosened and when rain falls and travels as surface run-off, it takes with it soil that washes into the river and silts up the river, displacing water.

3.11 Effects of floods

The effects of flooding can include the following:

- Buildings being washed away or damaged
- People and animals being drowned
- Communications being damaged due to closed roads and impassable railways
- Crops being ruined (and the agricultural economy suffering)
- Insurance claims
- Drinking water being contaminated by sewage, leading to disease

A beneficial effect is that silt deposited by flooding rivers provides fertile soil for farming.

3.12 Flood control and prevention

Various things can be done to prevent and control flooding:

- The construction of dams to control the amount of water being discharged
- The construction of levees and dykes to contain water
- The straightening of meanders to enable flood water to escape more quickly
- Afforestation to increase transpiration and infiltration (water will be removed from the river basin as it is taken up by the roots)
- Sandbagging to prevent the flooding of buildings
- Avoiding building on flood plains – flood plain zoning
- Ecological flooding – allowing meadows to flood to prevent flooding elsewhere
- Various types of sea defence, such as sea walls, groynes, rip rap or offshore breakers, to reduce the risk of coastal flooding

3.13 Case study – Boscastle floods, 2004

Flash floods occurred in the valleys of the River Valency and the River Jordan on 16 August 2004.

Causes

Physical

The village of Boscastle is situated:

- in a narrow valley with interlocking spurs which acted like a funnel
- in a steep valley which encouraged rapid run-off
- on a flat flood plain
- in an area where the soil is impermeable clay which does not allow much infiltration.

Climatic

- 185 mm of rain fell in five hours.
- The soil was saturated from recent rainfall so no more rain could infiltrate.
- The collision of winds on a very warm day caused the excessive rainfall. (The air mass from the south met the air mass from the south-west and converged on Bodmin which led to towering cumulonimbus clouds. The air was very unstable and the clouds were up to 10 km high.)

Human

- The natural channel of the river had been walled (for the construction of the B3263 and a pedestrian area) which prevented it from adjusting to a variation in the discharge of water.
- The village had been built on a flat flood plain.
- There was no flood control system.
- Cars, trees and boulders became stuck under the bridge and created a temporary dam which caused the water to build up behind it.
- The sewers and drainage systems were old and small in capacity; they broke and the water that was in them took an overland route.

Effects

- 50 cars were swept into the harbour.
- The bridge was washed away and roads were submerged under 2.75 m of water, making communication difficult.
- The sewerage system burst.
- For health and safety reasons Boscastle was declared inaccessible.
- The Museum of Witchcraft lost 50 per cent of its artefacts.
- Four buildings were demolished and 58 flooded and the High Street was badly damaged.
- The visitors' centre, a clothes shop and two gift shops were badly damaged.
- The youth hostel was flooded.
- People were in shock and there was concern about hypothermia or being swept away.
- There was no power in the village for some time. (An emergency generator had to be flown in.)
- 90 per cent of the economy in Boscastle is based on tourism and there were still three weeks of the summer holidays left; twenty accommodation providers were shut.
- Visitors whose cars had been washed away were not able to leave.

Responses

- A speedy, well-co-ordinated and well-resourced rescue operation ensured that remarkably there was no loss of life. Even by the standards of developed countries, this was outstanding and a tribute to Britain's rescue services.
- Emergency workers rescued residents and holiday-makers from a 32 km stretch of the north Cornwall coast.
- Hundreds were evacuated from homes, rooftops (120 from rooftops), trees and vehicles.
- Seven helicopters from the Coastguards, Royal Navy and RAF were used.
- People took emergency shelter in The Rectory, which is on high ground.
- The village was cordoned off by building inspectors for the clean-up operation.
- People dug out guttering and removed rubble so that the water could flow away.
- Sandbagging was used as a form of defence.
- People came to see the catastrophe.
- Prince Charles and the Deputy Prime Minister at the time, John Prescott, came to see the damage.
- There was a church service to give thanks that no one had died.
- The repairs were very costly and time consuming.
- There was a huge fund-raising effort to help rebuild the village.
- Insurance is now costlier in Boscastle.

Flood control and prevention

- The Environment Agency carried out a major investigation.
- A £2 million grant was given to Boscastle to help with flood prevention.
- No more schools or old people's homes are to be built in the valley.
- The Environment Agency removed debris from upstream, which meant there was more room for the water to flow freely under the new bridge.
- A flood defence system (building a flood wall, widening the River Jordan, raising car parks, removing bridges and using relief channels) was planned and is now complete. This also included building a high-arched single-span bridge which would not impede flood water and debris.

3.14 Case study – Flooding in Bangladesh, 2012

Floods occur each year in Bangladesh and are sometimes very serious, causing much loss of life. In 1998, for example, two-thirds of the country was covered in water, 30 million were made homeless and over 1000 people died. The majority of Bangladesh's 140 million inhabitants live on the flood plains of the Ganges and Brahmaputra and they need the floods to enable them to grow rice and jute. The floods also deposit silt, which makes the soil fertile. However, the inundation (flooding) is often so intense that lives and crops are ruined. Bangladesh suffers from two types of flood: river flooding and storm surges (coastal flooding) from the Bay of Bengal.

Bangladesh can get more rain in four months than London gets in two years!

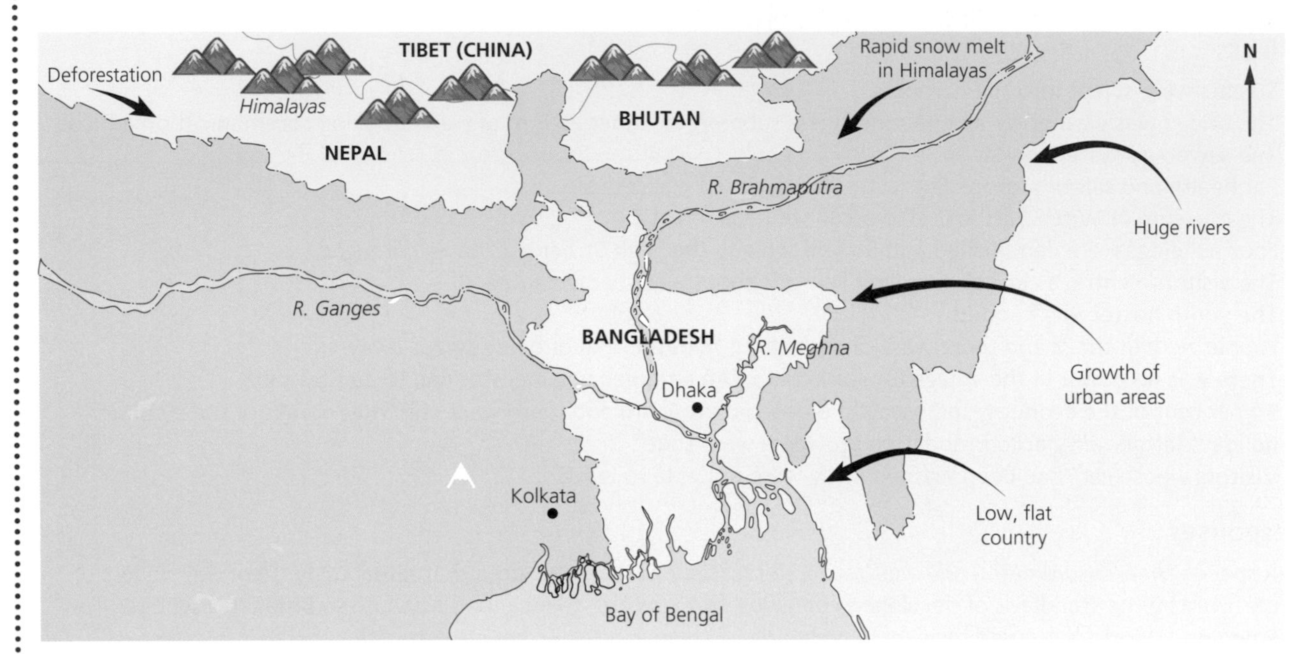

■ Figure 3.16: Location and causes of the flood in Bangladesh

Causes

Climatic

- Bangladesh has a monsoon climate, receiving between 1800 mm and 2600 mm of rainfall per year. However, 80 per cent of this rainfall takes place in four months (June to September). On one day in June 2012, Chittagong received 40 cm of rain in a single 12-hour period.
- High temperatures from June to September cause ice and snow to melt in the Himalayas where the Ganges and Brahmaputra have their sources and tributaries.
- Tropical cyclones which are funnelled up the Bay of Bengal make sea levels rise and stop the river flood water escaping. As the land becomes shallower, the water builds up to form a surge up to 6 metres in height.

Physical

- Half the country lies less than 6 metres above sea level.
- Most of the population live on the silt deposited by the Ganges and Brahmaputra Rivers, which forms a delta. However, the continuous deposition of silt tends to block the main channels and raise the height of river beds, making severe floods more likely.
- Once rivers overflow their banks, the water can spread a vast distance across the flat delta flood plain.
- High tides in the Bay of Bengal prevent the flood water from escaping.

Human

- Global warming is causing glaciers in the Himalayas to melt and the sea level of the Bay of Bengal to rise.
- Urbanisation on the delta flood plain has led to more run-off and a shorter lag time (the time between maximum rainfall and maximum discharge in the river).
- Deforestation in the upper course of the river (Nepal) has led to more run-off which allows more sediment to build up, which leads to a higher risk of flooding.

Effects

Environmental

- The flood was so deep in places that only the tops of roofs and trees could be seen.
- Landslides were caused by the soil becoming saturated.
- 'Char' areas (low flat land made up of deposited silt) were destroyed.
- The deposits left by this severe flood were infertile sand rather than silt. When the water receded, the land was infertile.

Social

- At least 139 people were killed.
- The city of Chittagong was deluged.
- In a developing country many people are too poor to own a telephone or TV and so they did not get advance warning to escape.
- There were food shortages due to the death of livestock and submerged crops (rice, jute and sugar).
- Infrastructure, such as roads, railways and bridges, was destroyed.
- 360 000 homes were destroyed or damaged and 50 778 people were evacuated.
- Some parts of the country were without electricity for several weeks.
- Flood water caused some wells to become polluted.
- As people evacuated to higher ground, crowding aided the spread of dysentery, cholera and diarrhoea.
- Hospitals were crowded.

Responses

- Food grain was imported.
- Medical care and water purification tablets were provided in treatment centres and by mobile teams.
- Newspapers gave advice on how to avoid drinking dirty water.
- Rice and water was given out by the Save the Children Fund and Oxfam.
- After the extremely serious flooding in 1989, the Flood Action Plan (costing over $650 million) was proposed by rich countries to be funded by the World Bank. The plan included:
  - building large embankments
  - early flood warning systems
  - flood shelters made of concrete and built on stilts
  - dams
  - co-ordinated after-care such as food, water, tents and medicine and seed for next year's crop
  - reducing deforestation in Nepal.

Some of these proposals have been carried out. However, there has been some criticism of this plan and whether it is sustainable, as embankments restrict the channel, affect the fishing industry and increase the height of the rivers. Dams are also expensive and could lead to debt.

3.15 Case study – Hurricane Katrina (coastal flooding), 2005

Hurricane Katrina hit the Florida coast on 25 August 2005. She veered inland towards Louisiana making landfall at Grand Isle (90 km south of New Orleans). She then hit New Orleans, Biloxi, Mobile and Jackson. Katrina was a Force 5 hurricane on the Saffir-Simpson scale (which means the wind speeds were faster than 249 km/h). A storm surge (wave) built under the hurricane and flooded the area when the hurricane reached land.

Causes

Climatic

- The hurricane formed when a cluster of thunderstorms drifted over the warm Caribbean Sea.
- The warm air from the ocean and the storm combined and rose – this created low pressure.
- Trade winds blowing in opposing directions and the Coriolis force (a directional pull caused by the spin of the Earth) caused the storm to spin.
- The rising warm air caused pressure to decrease at higher altitudes.
- The air rose faster and faster to fill this low pressure, drawing more warm air off the sea and sucking cooler, drier air downwards.
- The storm moved over the ocean and picked up more warm, moist air. Wind speeds started to pick up as more air was sucked into the low-pressure centre.
- There was an eye of calm winds surrounded by a spinning vortex of high winds and heavy rain.
- A large dome of water was created under the hurricane, which turned into a storm surge when it hit land.

Human

- New Orleans is built in a bowl – the Mississippi had been controlled with levees and dams for years, and New Orleans was only in existence because of these.
- The city was suffering from subsidence as groundwater was extracted from underground and the coastline was disappearing. Louisiana has the highest rate of erosion in North America.

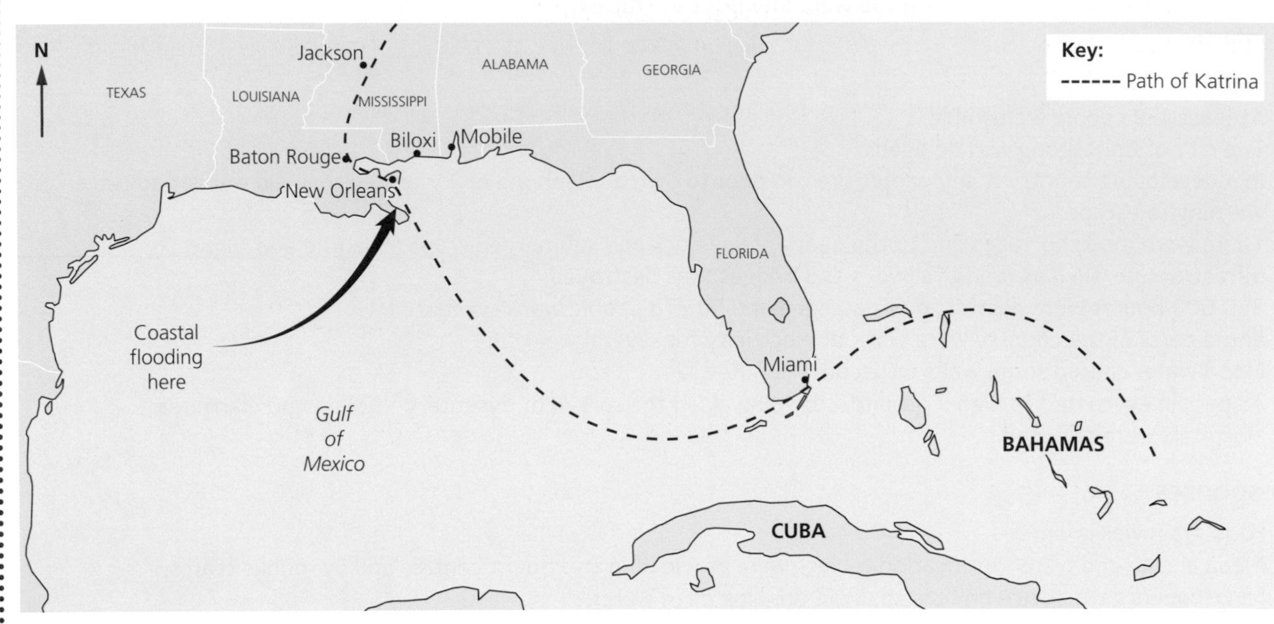

■ Figure 3.17: Path of Hurricane Katrina

Effects

Environmental

- A surge of 9 metres in Biloxi covered the land with water.
- Protective coastal mangroves were destroyed.
- The poorly maintained levees on the Mississippi broke, inundating New Orleans with flood water (80 per cent of the city was inundated).
- Flood water and fires destroyed many buildings.
- Trees and vegetation were destroyed by the flood.
- Fresh water was contaminated with salt water.

Social

- 1300 people were killed, mainly in New Orleans.
- There were 500 000 refugees.
- The flood water was polluted with oil and infested with snakes.
- Those who could not evacuate were left with a lack of water, food, power and fuel. They could not recharge their mobile phones and therefore could not use them to communicate.
- The disaster revealed the poverty of New Orleans where, for example, 50 per cent of children were on welfare. Afro-American leaders became angry that this section of society was left behind.
- Poverty resulted in looting for food and the army's violent response came more quickly than aid.
- Many took refuge in the Superdome despite the fact that there was no proper sanitation or proper supplies; it became crowded and overheated.

Economic

- Work at oil refineries and platforms was halted.
- Casinos and businesses were looted.
- 400 000 homes and businesses were without power in Alabama.
- Huge blasts from a chemical plant rocked the city on 2 September.
- Many left their businesses and have not returned, as they have been forced to start a new life elsewhere.

Responses

- 'Katrina was a national failure.' (Congressional Report, February 2006)
- Aid was delayed by five days – the Director of the Federal Emergency Management Agency (FEMA), Michael Brown, was dismissed from his post a week after the hurricane.
- The evacuation drill consisted of saying 'leave town'. Many could not speak English, could not afford to leave or were too ill to leave. Disaster drills did not consider the levee break.
- President Bush had poor and incomplete advice, and a late decision to carry out a compulsory evacuation led to deaths and prolonged suffering.
- There was no warning and no buses were provided for evacuation.
- Helicopters dropped sandbags into the breach in the 17th Street Canal and earthmovers built a causeway allowing trucks to bring stones to repair levees.
- A temporary steel barrier was built at the mouth of the canal to seal it from Lake Pontchartrain.
- Areas were then pumped free of water.
- The Red Cross served 995 000 meals in one day alone.
- The congressional committee spent months planning how to rebuild and revitalise housing, business and transport, and how levees and flood defences could be improved to prevent large-scale flooding in the future.
- Bush promised $3.1 billion towards repairs and improvements.
- The US Army Corps planned temporary repairs in time for the next hurricane season.
- The Mississippi River Gulf Outlet, built in the 1960s, allowed ships easy access to the Gulf of Mexico via the Port of New Orleans. However, the channel increased from 91 m to 914 m wide due to deliberate widening and natural action. The storm surge of Katrina passed up this channel. The channel was closed in 2009.

Other facts

- FEMA suffered from a lack of trained and experienced personnel.
- Computer models predicted the levee failure, so the authorities should have been prepared.
- The impact of global warming causing sea levels to rise will make flooding an even greater threat in the future.
- Some believe that the chaos of Katrina has exposed deep divisions in New Orleans and US society. (Congressman Elijah Cummings said, 'We cannot allow it to be said by history that the difference between those who lived and ... died ... was nothing more than poverty, age or skin colour.'

3.16 Case study – Flooding in Tewksbury, 2007

Tewkesbury is situated in Gloucestershire. It was the worst affected part of the country when the floods hit in July 2007.

Causes

- The rivers Severn and Avon meet in the centre of Tewksbury. Both of these rivers overflowed their banks.
- Soil in the drainage basin was already saturated by previous rainfall.
- Summer 2007 was the wettest in England and Wales since 1766. This was caused by a low pressure system or depression over the UK. A jet stream which was situated further south than normal pushed this depression south to sit over the UK.
- 119 mm of rain fell on Friday 20 July. This is the amount of rainfall usually expected over two months.
- There was little sunshine and therefore evaporation rates were low. Combined with the high rainfall, this led to flooding.
- Building on flood plains in Tewkesbury meant that there was a great deal of concrete so infiltration was decreased and surface run-off increased.
- No flood defences had been built.

Effects

- 1800 homes and 300 businesses were damaged by flood water.
- 95 per cent of homes were without water for a period of time.
- Three people died; two died due to fumes from a diesel water pump.
- Four of the main roads to access Tewksbury were impassable.
- 200 families still had no home to return to a year after the floods.
- A water sewage treatment works was flooded and this contaminated drinking water for 1400 homes.
- Crops which had been affected by sewage had to be destroyed.
- Flooding cost the local council £140 million.

Human responses

- RAF helicopters were used for rescue.
- The Army provided food for the cut-off town of Upton-on-Severn.
- £800 million was spent on flood defence.
- Council offices were open 24 hours for six days.
- Portaloos were distributed by the council.
- A temporary doctors' surgery was created for one week.
- 25 000 sand bags were used.
- 180 000 insurance claims were processed.

Revision tip

Whichever case study you have chosen it is a good idea to make a mind map of the details. Put the name of the case study in the middle with the date on which it occurred and then add sticks for causes, effects (split into environmental, social and economic) and human responses.

★ Make sure you know

★ The different rock types

★ The different types of weathering

★ The features of a river basin

★ The different river processes

★ The features of the upper and lower courses of a river

★ The features and causes of coastal erosion

★ The main features of coastal transportation and deposition

★ The causes and effects of and human response to a specific flood

Test yourself

Before moving on to the next chapter, make sure you can answer the following questions. Answers are near the back of the book.

1 Using some of the words given below, write out the following paragraph, filling in the gaps.

transports outside speed inside reduces river cliff increases river beach slip-off slope delta

Velocity is the __________ of the water. Deposition is the 'dumping' of a load when the river's velocity __________. Load is the material which the river __________. Load can be deposited by rivers at their mouth; the feature formed is called a __________. It is also deposited on the __________ bend of a meander; the feature formed is called a __________.

2 Copy and complete the table to show how the shape and size of a river's load changes from upper course to lower course.

	Upper course	Lower course
Size of load		
Shape of load		
Main methods of transportation		

3 To which type of weathering will the following be most prone?

(a) Rocks in a tropical rainforest

___

(b) Mountainous areas

___

(c) Rocks in a desert

___

4 Match the definition to the term and write them down. (The first one is done for you.) Note: you will have come across some of these terms in Chapter 2.

River basin – an area of land drained by a river and its tributaries

River basin	the movement of water over the surface of the land back to the sea
Watershed	rocks that allow water to pass through
Source	when the river's load collides and breaks into smaller pieces
Mouth	the downwards movement of water through tiny pores in the soil
Permeable	the movement of water through the soil back to the sea
Impermeable	the loss of moisture to the air from plants
Evaporation	the amount of water that passes a given point at a given time, measured in cumecs (cubic metres per second)
Transpiration	an area of land drained by a river and its tributaries
Through flow	the start of a river
Ground water storage	where a river meets the sea
Infiltration	water stored in rocks below the ground
Surface run-off	the material that a river carries
River discharge	a type of erosion caused by the force of the water breaking particles of rock from the river bank
Load	rocks that do not allow water to pass through
Attrition	the loss of water to the air when the water has turned into water vapour
Corrosion	a type of erosion caused by the acids in the river dissolving the rocks
Hydraulic action	the boundary of the river basin, usually marked by a ridge of high land

5 (a) Which of the following is a process that involves waves hitting cliffs and eroding them?

hydraulic action **deposition** **longshore drift** **saltation**

(b) Longshore drift can create a:

waterfall **stack** **stump** **spit**

(c) A stump is an eroded:

cave **stack** **spit** **meander**

6 Draw a diagram to explain the process of longshore drift.

7 **(a)** For an area you have studied, explain why there was a flood.

____

(b) Describe the effects of the flood.

____

(c) How can floods be controlled?

____

8 Write the definitions of these words and phrases you need to know, and then ask someone to check them.

abrasion

____

____

drainage basin

____

____

arch

____

____

erosion

____

____

attrition

____

____

estuary

____

____

bay

____

____

fault

____

____

beach

____

____

fetch

____

____

confluence

____

____

flood plain

____

____

corrosion

____

____

gorge

____

____

delta

____

____

headland

____

____

hydraulic action

hydro-electric power

igneous rock

impermeable

joint

levee

longshore drift

lower course

meander

metamorphic rock

middle course

mouth

ox-bow lake

permeable

plunge pool

porous

rapids

reservoir

river basin

river cliff

saltation

scree

sedimentary rock

slip-off slope (river beach)

solution

transportation

source

tributary

spit

upper course

stack

waterfall

suspension

water table

swash

weathering

4 Population and settlement

4.1 Population statistics, terminology and factors that affect density

Population figures are constantly changing. Find out what the current population figures are and fill them in on a copy of this table.

Place	Population
World	
China	
India	
UK	
London	

Population densities

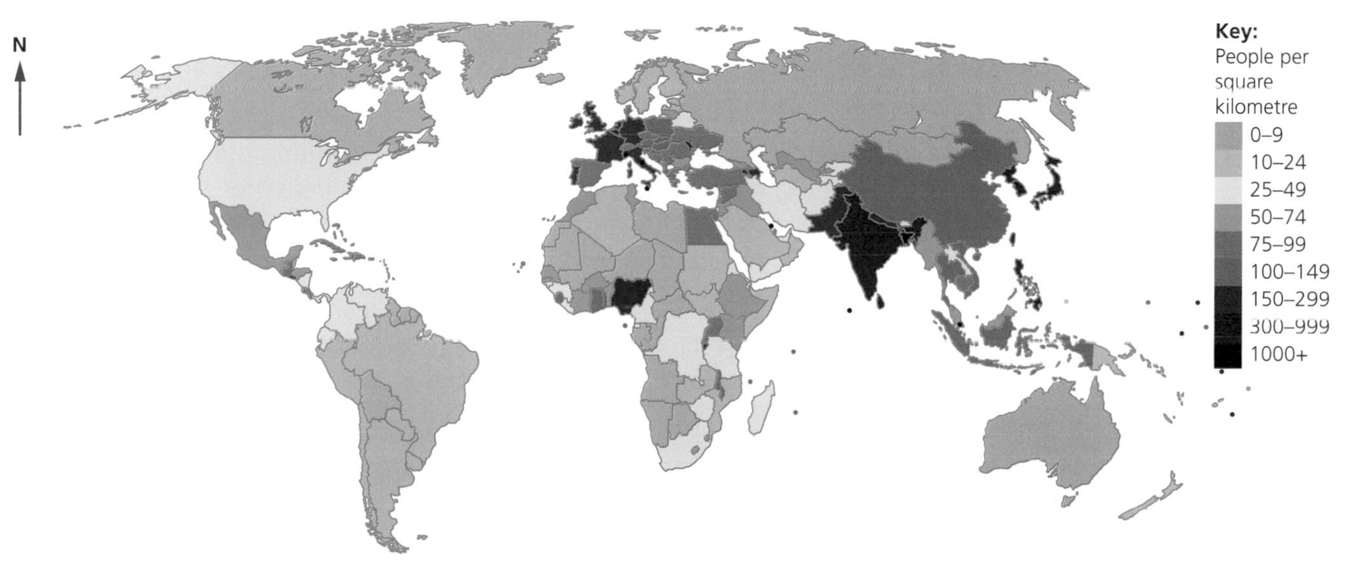

■ Figure 4.1: Choropleth map showing world population densities (2014)

Population densities (2014)

World 53 people per square kilometre (land only and excluding Antarctica)

Bangladesh 1034 people per square kilometre

UK 262 people per square kilometre

Australia 3 people per square kilometre

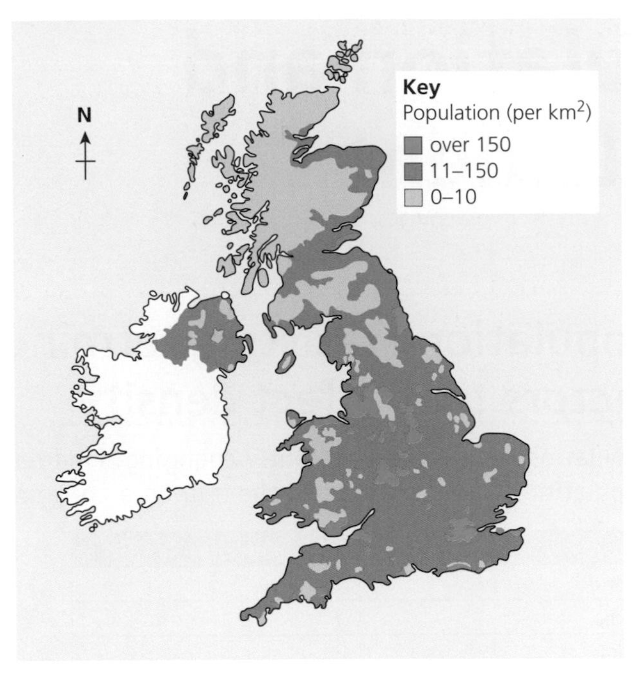

■ Figure 4.2: Population distribution in the United Kingdom

Factors that affect population density

Climate:

- Climates which are temperate are ideal for settling.
- Places which are extremely hot, cold or dry tend to have lower population densities, for example the Sahara Desert is very hot and dry and therefore very few people live there.
- Mountainous areas of the UK, which are wetter, windier and colder than the rest of the UK, have comparably lower population densities than the places with more favourable climates.

Relief of the land:

- Places which are flat or have gentle relief are best for settling because this usually provides a temperate climate and is easier for the construction of infrastructure. Therefore the Himalayas has a low population density.
- Mountainous areas of the UK, such as the Grampians or Snowdonia, have lower population densities than flatter areas such as the southeast of England.

Fertility of soil:

- Areas where soil is fertile provide for productive agriculture. Therefore population densities are greater in these areas.
- Fertile areas include land near rivers and at the base of volcanoes.
- Bangladesh, on the Ganges floodplain, has a high population density.
- Fig 4.2 shows that East Anglia, with its fertile boulder clay soil, has a reasonably high population density.

Availability of natural resources:

- Countries or areas which are rich in resources such as oil, coal and precious minerals are likely to have high population densities.

Perceived likelihood of natural disaster:

- Countries which are away from tectonic plate boundaries and are less likely to experience other natural disasters such as hurricanes and floods are likely to be more densely populated. For example the UK is not likely to experience severe tectonic disasters or hurricanes and this is one reason why it has a high population density.

Stability of government:

- Countries with unstable governments often experience high levels of emigration. For example the Central African Republic has experienced many coups and has low population density.

Job opportunities:

- Areas with many job opportunities are likely to have high population densities. For example the southeast of the UK has a much higher population density than the rest of the UK, and this is partly due to the fact that London provides a multitude of jobs and high paid jobs.

Revision tip

You could draw a spider diagram of all of the factors that affect population density. Write 'population density' in the centre and then add sticks with all of the factors coming off the centre. You should use pictures and words rather than sentences. Remember to use colour effectively.

Population terminology

- The birth rate is the number of births per thousand of the population per year.
- The death rate is the number of deaths per thousand of the population per year.
- The rate of natural increase is the difference between the birth and death rates (RNI = birth rate – death rate).
- Migration is the movement of people from one area or country to another to find work or a better standard of living.

4.2 Causes of the UK's increase in population since 1945

The UK's population rose by over 10 million between 1945 and 2000. It is expected to grow by 6.1 million to 65.7 million by 2031 and then reach a peak at 67 million by 2051.

Death rate and birth rate

Death rate has been low in the twentieth century due to great improvements in medical care and nutrition. Currently in the UK it is about nine (deaths on average per 1000 people each year).

Birth rate has reduced considerably. This has been due to:

- the increasing ease of access to contraception
- agriculture becoming mechanised so fewer children are required to work the land
- infant mortality rates lowering
- the desire for material possessions taking over from the desire for large families
- equality of women meaning that many have pursued careers rather than staying at home.

However, birth rate in the UK is higher than many countries in the rest of Europe; currently it is about twelve (births per 1000 people each year).

With a birth rate of twelve and a death rate of nine, the current rate of natural increase (difference between birth rate and death rate) in the UK is three.

Immigration

There has been a considerable amount of immigration into the UK. This has increased the population considerably.

- Many people living in former colonies of the British Empire such as in India, Bangladesh, Pakistan, the Caribbean, South Africa, Kenya and Hong Kong have arrived in the UK.
- Many others have come as asylum seekers seeking protection as refugees from war torn countries such as Afghanistan, Sudan, Iran and Zimbabwe.
- There has been much immigration from countries within the EU. In 2004, Poland and seven other Eastern European countries joined the EU and this led to much immigration to the UK.
- In 2005, about 565 000 migrants arrived in the UK and only 380 000 people left the UK to live abroad. 57 000 of these immigrants were from Poland.

Revision tip

You could draw some pictures to help you remember all of the reasons why the UK's population has increased since 1945.

4.3 Functions of a settlement

A settlement is a place where people live.

The functions of a settlement are the things that happen there. A settlement may have more than one function and these may change over time.

As well as a residential function, meaning that people live there, a settlement may have a number of other functions:

- An industrial function means that factories are located there. These now tend to be in out-of-town locations in the outer suburbs.
- A commercial function means that shopping facilities are located there. This may take the form of shopping centres, cinemas, leisure centres, etc.
- A service function may include schools, hospitals, libraries, etc.
- A tourism function will be of a particular kind, depending on the type of settlement.
- An administrative function means that local government has offices there, from which it runs public services.

4.4 Reasons for the site or situation of a settlement

- The site of a settlement is its exact physical location.
- The situation of a settlement is its setting in relation to surrounding features.

Most settlements grew up in ancient times, before motorways and tourism. Early settlers would have considered the following factors:

- Relief – settlers would choose an area that was high enough to be safe from flooding but low enough to be sheltered from winds.
- Transport – settlers would choose a site near the fording or bridging point of a river, at a crossroads (originally tracks rather than roads) or near the coast as this made travel more easy.
- Soil – settlers would choose areas with deeper and more fertile soil as this is better for agriculture.
- Water supply – settlers would choose a site near a river, spring or well as they needed water for cooking, cleaning and drinking.
- Wood – settlers would choose to settle near woodland as they used wood for building and fuel.
- Defence – settlers would choose hilltops, marshes and meander bends as these sites were easier to defend.

Revision tip

Try drawing a picture of the perfect village in a location that provides all of these factors!

4.5 Settlement hierarchy

Settlements can be ranked in order – a hierarchy. The order within the hierarchy is decided by population, area, and range and number of services.

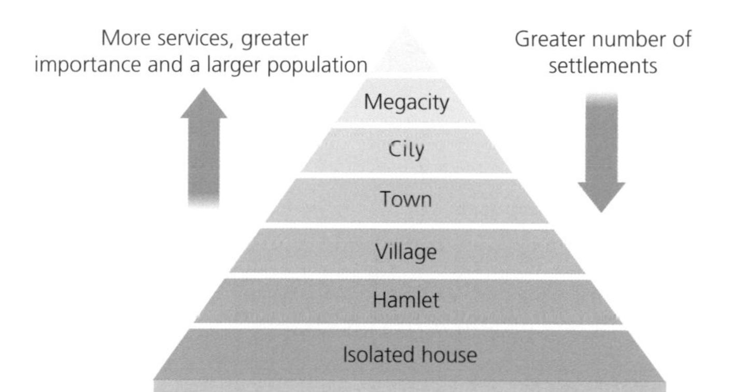

■ Figure 4.3: The settlement hierarchy

The larger the settlement, the more services it will have.

Settlement	Services	Approximate population
Hamlet	Perhaps none	Less than 100
Village	Church, public house, convenience shop (although many are disappearing), primary school	100–2500 people
Town	Several shops, churches, secondary school, dentist, bank, small hospital (although fewer and fewer exist in towns)	2500–100 000 people
City	Cathedral, large railway station, large shopping centre, large hospital, specialist shops, museum	More than 100 000
Megacity	Cathedral, large railway station, large shopping centre, large hospital, specialist shops, museum	More than 10 million

4.6 Settlement patterns

Settlements develop in a pattern. The main settlement patterns are:

- linear
- dispersed
- nucleated
- planned.

Many settlements contain a mixture of these shapes.

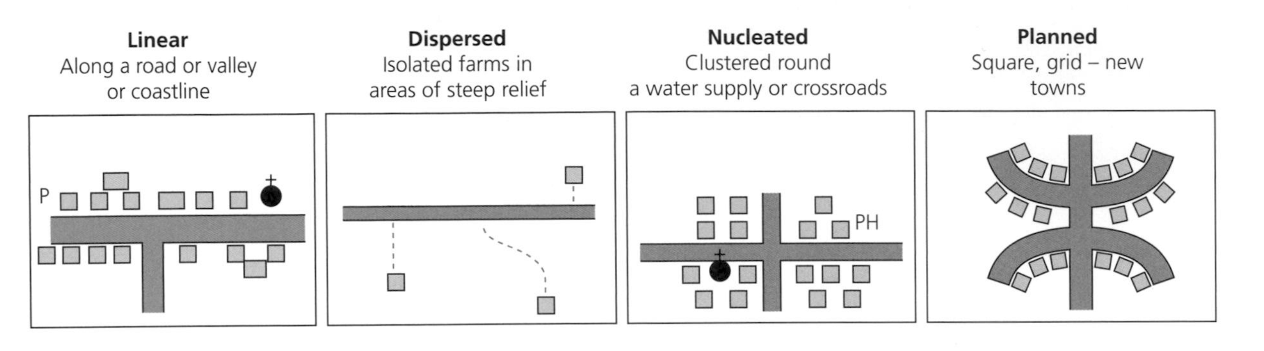

■ Figure 4.4: Settlement patterns

Linear settlements (also known as ribbon settlements) developed as houses were built along transport routes. As transport improved in Britain in the 1920s, people could live further from work, and urban sprawl occurred together with linear settlements along new transport routes. Green belts (where planning permission is limited) were introduced to control urban sprawl.

Revision tip

Try thinking of hand gestures to represent linear, dispersed, nucleated and planned. Once you and your friends have decided on the gestures you could test each other.

Revision tip

On an OS map try to find as many linear, planned, dispersed and nucleated settlements as possible. Then work out why you think each one has grown into this shape.

4.7 Reasons why some settlements grow and others do not

Some settlements grow bigger than others for the following reasons:

- Lack of competition – there are no other big settlements nearby.
- Villages or small towns that are near cities may grow bigger as commuters come to live in them. This is called suburbanisation.
- If a new industry locates in a settlement it will create jobs, which will encourage people to migrate to this place. On the other hand, a lack of job opportunities will encourage people to leave; for example, in rural areas there are now few jobs available in agriculture or mining.
- Accessibility and flat land encourage growth whereas marshy or steep land can inhibit the growth of a settlement.

4.8 Case study – Environmentally sensitive planning: East Village in Queen Elizabeth Olympic Park

The 2012 Athletes Village has been recycled to create a new housing development called East Village. The houses are a mix of private, affordable rental and shared ownership. A total of 2818 homes have been created, with many services including a school for 3–18 year olds called Chobham Academy and a nursery. East Village is located within the 27 acres of Queen Elizabeth Park. It has 30 local independent shops and cafés and its own health facilities, the Sir Ludwig Guttmann Health and Wellbeing Centre. There are also fantastic facilities for children including Tumbling Bay playground and Wild Kingdom playground. Westfield Stratford City's array of shops is easily accessible from East Village. All London airports can be reached within an hour and the West End is only 20 minutes away.

How has sustainability been considered?

The development was planned with sustainability in mind: sustainable design, sustainable construction, sustainable materials, and sustainable homes.

- East Village is extremely well served by public transport, which not only eases car congestion but also reduces carbon dioxide emissions. There are 9 rail links, 195 trains per hour, 15 local bus services and the newest and safest cycle routes. The Thames Clipper can also be easily caught from nearby Greenwich. Stratford International station and Stratford Docklands Light Railway (DLR) station are a five-minute walk away. On the fringes of East Village is Lee Valley Regional Park which contains lots of cycle tracks and 42 km of canal walks.
- Only certified timber has been used in the construction of the village.
- During construction, 4000 local people were trained and worked on the building site.
- 183 000 tonnes of carbon will be saved annually by the insulation of the homes.
- 90 per cent of construction waste was directed away from landfill.
- Natural light has been incorporated into the design so that fewer lights have to be switched on inside the homes. Use of LED lights cuts carbon emissions by 5000 tonnes per year.
- Living green roofs have been planted on all buildings higher than 100 m. These absorb carbon dioxide, encourage wildlife and reduce noise.
- A biomass power station on site delivers heat and energy to the development.
- East Village has a water-recycling project. Water from gutters and roofs is reused to keep gardens looking good. Then it is filtered by the wetlands and bought back into the system. Grey water from showers and washing machines is reused for flushing loos. A third less water will be used in East Village per person than in the UK as a whole. Landscaping of the area has enabled surface run-off to be caught so that flooding risks are reduced.
- In May 2011 a rare Black Redstart was found to be nesting on site so construction halted until it had finished nesting.
- There is a community orchard which adds to the biodiversity of the area and also reduces the air miles of some food supplies.
- Wetlands, treetop walkways and Mirabelle Gardens are all areas of green space where wildlife, including Dunnocks and House Sparrows, is encouraged to thrive.

Revision tip

Whichever case study you have chosen as your sustainable housing project, it is a good idea to make a mind map of the details. Put the name of the case study in the middle and add a stick for each of the ways in which sustainability has been achieved. Remember to use colour to good effect and to use revision pictures and words rather than sentences.

Exam-style questions

Try these questions. Answers are near the back of the book.

4.1 Why are some countries densely populated? (3)
4.2 What is the current world population? (1)
4.3 What type of settlement would have a cathedral and more than one secondary school? (1)
4.4 What shape is a coastal settlement likely to have? (1)

★ Make sure you know

- ★ Population statistics for the world and the UK
- ★ Causes for the variations in population density
- ★ Why one country's population has risen or fallen
- ★ The different functions of a settlement
- ★ The reasons why locations were chosen for particular settlements
- ★ The settlement hierarchy and different settlement patterns
- ★ How a housing project is being or has been developed in a sustainable manner

Test yourself

Before moving on to the next chapter, make sure you can answer the following questions. Answers are near the back of the book.

1 Why has the UK's population increased?

____

____

2 Why may a settlement be in a nucleated shape?

____

____

3 Describe how the characteristics of a settlement change as you go up the settlement hierarchy.

____

____

4 Write the definitions of these words and phrases you need to know, and then ask someone to check them.

birth rate

____

____

bypass

____

____

brownfield site

____

____

CBD (central business district)

____

____

continent

____

____

death rate

____

____

desert

____

____

detached

____

____

dispersed

____

____

distribution

____

____

ethnic group

____

____

function

____

____

greenfield site

____

____

hierarchy

____

____

high order settlement

____

____

land use

____

____

life expectancy

____

____

linear

____

____

low order settlement

____

____

migration

____

____

natural increase

____

____

nucleated

____

____

pull factor

____

____

push factor

____

____

resource

____

____

retail

____

____

rural

____

____

semi detached

____

____

settlement	suburb
settlement pattern	terraced
site	urban
situation	urbanisation
social	

Revision tip

Try making flash cards to help you remember the words you need to know. Write the word on one side of the card and the meaning or a picture to represent the meaning on the other. Use them to test your friends.

5 Transport and industry

5.1 Advantages and disadvantages of different modes of transport

Mode of transport	Advantages	Disadvantages
Car/road	• Convenient • Short journey time unless in congestion	• Carbon dioxide released • Cost • Danger to self and others • Noise • Destruction of environment to create roads
Foot	• Healthy • Free • Only option in many developing countries	• Longer journey time • Uncomfortable if very hot or cold
Bus/road	• Less carbon monoxide released per person than car • Can use bus lanes • Cheaper than using a car	• Cannot go at exact time of choice • Can suffer from congestion
Boat/sea	• Can carry heavy freight a long distance cheaply	• Takes longer than air travel
Plane/air	• Quick • Safer than road and rail • Good for transporting light, valuable freight long distances	• Expensive for people and freight • Carbon dioxide released • Noisy for people who live on flight path • Have to use another mode of transport to get to airport and to destination
Bicycle	• Healthy • Cheap	• Danger, reduced by use of cycle lanes • Cannot carry freight
Lorry/road	• Most efficient for carrying freight short distances to specific points; most freight is transported this way in the UK	• Carbon dioxide released • Noisy
Rail	• Can carry freight medium to long distances reasonably cheaply	• Can be expensive for passengers • Have to use another mode of transport to get to exact destination • Destruction of natural habitats to create railway tracks • Noise • Vibrations of trains can disturb foundations of buildings

5.2 Value of transport routes

- People need to get to work (commute) and children need to get to school using transport infrastructure.
- People who live in rural areas need to access urban areas for shops and services.
- People use transport for leisure/tourism and for socialising.

- Industry needs transport for accessing raw materials, for getting the finished product to the market and for labour to get to work.
- Quality of life is improved for people who have good access to transport infrastructure and it also brings an increase in the value of property.

5.3 Case study – High Speed 2

High Speed 2 (HS2) is a proposed train, which will link London to cities further north. This plan will increase passenger capacity. Phase one is to create a line between London and the West Midlands by 2026. Trains will run at speeds of up to 400 km per hour and 14 trains will run per hour. Phase two will extend the line in two branches from Birmingham, one branch running to Leeds and one branch to Manchester.

There are many arguments both for and against this project. Some of these are listed below:

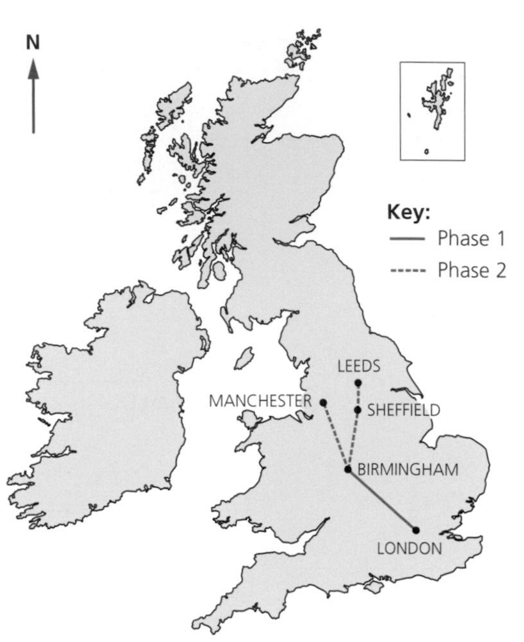

Figure 5.1: Route of HS2

For

- High speed rail will provide a green, safe and efficient form of transport.
- Journey times between London and Birmingham will only be one hour.
- The project will be a major boost to businesses and the economy in Birmingham and in London.
- Carbon dioxide emissions will be reduced as less people will be travelling by car or plane.
- There will be less congestion and fewer accidents on the roads, especially in some gridlocked cities.
- Jobs will be provided building and running HS2.
- Hopefully it will lessen the north/south divide in terms of job availability, wealth and house prices.
- Costs will be spread out over 15–20 years, so it is affordable.
- Trees will be planted along the tracks and there will be 'green tunnels'.
- Property prices at either end of the line will increase considerably.

Against

- The cost of the project is escalating and is now forecast to be £42.6 billion.
- As an alternative, train length could be increased on existing routes, therefore saving money and environmental destruction.
- This project will only benefit a small section of the population.
- Business people can work on a train using Wifi; this brings into question the necessity to have a high speed train.
- There will be a great deal of environmental destruction to create the track and electricity pylons, especially to the green belt of London and the Chilterns.
- It is not entirely definite that there will be adequate demand for this service.
- There will most certainly be noise and visual pollution associated with the project leading to a decrease in living standards for many and a permanent change in the character of many towns and villages.
- The price of homes will decrease for those who live near the line by 25–35 per cent. Homes on the proposed line will be compulsory purchased.
- The town of Amersham and the villages of Great Missenden, Culworth and Eydon will all be negatively affected.
- Tourism will be negatively affected in areas such as the Chilterns.
- 83 hectares of woodland will be adversely affected.
- The migration patterns of some wildlife will be affected.
- The tunnel at Amersham could disrupt the balance of the River Misbourne and may cause a lake to dry up.
- Tunnels could also affect underground reservoirs of water which could disrupt water supplies for north Oxfordshire and Buckinghamshire.

5.4 Case study – Heathrow third runway

Many business people think that the UK needs to expand its airport capacity in the South East so that the economy can grow. The quickest and easiest way to do this is to include a third runway and sixth terminal at Heathrow. Many environmentalists are not in favour of this and many would prefer a new airport built in the Thames estuary or expansion at Gatwick or Stansted. Arguments for and against a third runway at Heathrow include the following:

For

- Passenger numbers are predicted to increase from 160 million/year in 1998 to over 400 million/year by 2020 so greater air travel capacity is needed.
- Without expansion, the UK will lose out on business competitiveness and tourism.
- Building a new airport in a different location, for instance on an island in the Thames, would take much longer than expanding Heathrow.
- Heathrow already has good transport links so expanding it is the cheapest way of creating additional capacity.
- High Speed 2 could be expanded from London offering a fast connection to Heathrow from Birmingham.
- 60 000 jobs would be provided by the creation of the third runway.
- This could increase Heathrow's efficiency and therefore reduce the emissions of planes waiting to land.

Against

- 725 000 people already live under the Heathrow flight path; it covers an area of high population density. Airport expansion would mean that even more homes would be affected by noise pollution.
- 700 homes would be destroyed and the village of Sipson would be wiped out.
- New roads would have to be built to access the runway and terminal; one of these will cut through Cherry Lane cemetery.
- Heathrow School will be demolished and The William Bird School will be at the end of the new runway.
- Eight grade 2 listed buildings and a church will be destroyed.
- Green belt land will be devastated, and agricultural land and ecosystems will be destroyed.
- Noise, air and visual pollution will increase in the area.
- Heathrow would become the biggest emitter of carbon dioxide in the country. The third runway could emit as much carbon dioxide per year as the whole of Kenya.
- Heathrow already has one of the highest numbers of international flights to key business centres (990 per week). This is many more than its rivals, such as Charles de Gaulle in Paris.
- With the growth of internet and tele conferencing it is possible that the number of business trips will not grow.
- Increasing capacity will increase carbon dioxide emissions. It would be better to reduce the number of short haul flights by increasing air fares and encouraging train travel, especially if HS2 goes ahead.
- Regional airports such as Manchester could be used to greater capacity.

Revision tip

You could have a debate with your family about whether or not you think HS2 or the third runway at Heathrow is a good idea. Or you could organise a debate with your friends, with some of you speaking for these new transport developments and some of you against.

5.5 Employment structure

The employment structure of a country is determined by the percentage of the workforce employed in each of the four types of activity:

- Primary industry, which extracts raw materials from the earth or sea, and employs, for example, farmers, miners, fishermen and forestry workers
- Secondary industry, or manufacturing industry, which makes raw materials into goods, and employs, for example, bakers and car-factory workers

- Tertiary industry, or service industry, which sells goods or provides a service, and employs, for example, doctors, lawyers and bankers
- Quaternary industry, or knowledge-based industry, such as research and development into high-tech goods, which employs, for example, research scientists

> **Revision tip**
> To help you remember these four types of industry, you could write down each one and then draw lots of pictures to represent the types of jobs in each category.

Employment structure in developing and developed countries

A developing country has a different employment structure to that of a developed country. In a developing country many more people work in primary activities, as subsistence farmers or in mining or fishing. In a developed country less people work in primary activities as mechanisation has meant that less workers are required in the agricultural and mining industries. Also, much food is imported into developed countries.

Secondary industry has increased in developing countries over the last 50 years as companies have taken advantage of the cheaper labour in developing countries to decrease their costs and increase their profits.

In developed countries many people work in tertiary activities as the population has disposable income to spend on leisure and services, and in shops. More taxes are collected by the government in developed countries, which means that more doctors, nurses, teachers and so on can be employed.

There are more people employed in quaternary industries in a developed country as more money is available for research and development.

5.6 Location of an industry

During the last century in the UK, traditional heavy industries, such as iron and steel, were located next to coalfields (for power supply), raw materials and railways. But industries in the UK today are generally high-tech and tend to be far less tied with regard to their location. However, company owners still have to consider the following when deciding where to locate:

- Labour force (where the workers live)
- Relief (whether the land is suitable for building on)
- Space (whether there is space available to build on)
- Market (where the people or firms that make up the market are located)
- Leisure facilities (whether facilities such as golf courses or health spas are nearby)
- Government grants (whether the government is offering money to locate in a certain area)
- Transport (proximity to motorways, and also railways and airports)

> **Revision tip**
> Try drawing a spider diagram with 'location of an industry' in the centre and then all the factors on sticks coming off the centre. The factors could be represented by pictures.

5.7 How economic activities can affect the environment

The environment is made up of the:

- landscape
- atmosphere of an area
- plants and animals that make that area their home (habitat).

Economic activities that can affect the environment

- Agriculture – Use of pesticides can kill certain insects and fertiliser ww into rivers can cause eutrophication (excessive growth of plant life). Habitats may be destroyed when land is cleared for cultivation. Some agricultural buildings cause visual pollution. Intensive animal rearing can cause bad smells in the area and noise pollution, and can be seen as cruel to the animals themselves.
- Tourism – Tourists can drop litter and cause footpath erosion. Tourist traffic releases carbon dioxide and some hotels cause visual pollution. Livestock can be scared by tourists' dogs.
- Mining – Visual pollution can be caused by mines and/or the waste material that is piled up near them. Quarries can cause dust and noise pollution. Habitats are destroyed to create quarries.
- Manufacturing – Noise pollution can occur and factories can cause visual pollution. Carbon dioxide is released by most manufacturing industry.
- Forestry – The natural environment is altered by the removal of natural vegetation and non-native trees being planted. Noise pollution can occur when trees are felled.

5.8 Sustainable economic development

The environment has been and continues to be damaged by human behaviour. Stewardship means looking after (or managing) resources in a sustainable way so that they exist for future generations.

It is therefore our responsibility to protect the environment for future generations by practising sustainable development or conservation. This means using resources or areas of land in such a way that they will not run out or be damaged for future generations. For example:

- Sustainable fishing would involve only catching breeds that are plentiful.
- Sustainable mining would include screening the quarry with trees, filling in the quarry and replanting when the work was finished or filling the quarry with water to make a reservoir which would create new habitats.
- Sustainable forestry would include planting as many trees as were cut down.
- Sustainable agriculture would include encouraging wildlife by planting strips of wild flowers around fields, replanting hedgerows and not using pesticides or chemical fertilisers.
- Developing a school in a sustainable way would mean encouraging children to walk to school if possible, turning lights and computers off when not in use and making sure recycling was taking place.

- A sustainable factory would minimise the amount of carbon dioxide used by using renewable energy
- Sustainable tourism would allow tourists to visit a place to boost the economy without causing any damage to the environment. Management techniques such as those shown in the table below could be used to make tourism sustainable.

Technique	Why is it needed	Effect on people	Effect on place
No rubbish bins	To stop littering near bins	They take litter home or drop litter	Area is litter free or more litter is dropped
Car parking areas	To stop people parking anywhere	They are encouraged to use the car park	Less vegetation damage and soil erosion elsewhere but car park could be classed as visual pollution
Logs at side of car park	To stop car park becoming larger	They park within the bays	Less vegetation damage and soil erosion
Screening of car park with trees	To ensure that car park does not cause visual pollution	They are attracted to the area as it looks beautiful	The area looks attractive
Footpaths	To stop vegetation trampling and soil erosion on other locations	They are encouraged to use paths	Vegetation trampling and soil erosion is limited to certain areas
Honey pot sites	To minimise environmental damage at other locations	They are encouraged to visit the honey pot sites	Honey pot sites suffer from negative environmental effects but other areas benefit
Visitor centre	To educate visitors	They understand the environment and conservation	The environment suffers less from human activity as people learn about the area but there are negative effects as it creates a honey pot site

5.9 Globalisation

Globalisation is the process by which companies become more distributed and ideas and lifestyles are increasingly spread and adopted around the world. Globalisation affects what we:

- eat; for example, we eat Indian and Mexican food
- wear; for example, people all around the world wear Nike trainers
- watch on TV; for example, we watch shows made in the USA, such as *The Simpsons*.

Globalisation has been aided by improvements in transport, communications and technology, which have meant that proximity to raw materials, energy supply and market are no longer as important for the location of a business. Goods can now be transported easily, energy supplies are available all over the world and markets have become global. Low labour costs have therefore become the most important factor affecting the location of a company.

Globalisation has led to the development of and also been driven by transnational corporations (TNCs) – also known as multinational corporations (MNCs) – which are companies with branches in many countries.

5.10 Case study – Nike: a multinational company

Nike is a transnational company.

- Nike employs 25000 people directly and one million others are involved in making, supplying and selling goods.
- In 2014, Nike had revenue of $25 billion.
- Nike sponsors Lionel Messi, Tiger Woods, Wayne Rooney and Ronaldo. These people are seen around the world wearing Nike clothing.

Reasons for choices of location

Nike headquarters (quaternary industry)

The company headquarters is located in Oregon, USA (a developed country) for the following reasons:

- The country has a high level of technology.
- Highly educated people are available therefore the headquarters can hire expertise.
- There is a highly developed transport infrastructure so workers can get to meetings easily.
- The USA influences the world in terms of the latest trends and fashion so the headquarters is at the core of this and can pick up ideas quickly.
- This is a prestigious location.

Nike retail (tertiary industry)

The goods made in the factories are sold in Nike shops, which are located mainly in southern and western Europe, but also in North America and Asia (and a very few in South America and Africa). Sales are highest in Canada, USA and Europe (developed countries).

- The main market is in developed countries where customers are more affluent.
- In developed countries people are more influenced by adverts they see on TV or computers and are more likely to buy goods with a recognisable label.

Nike manufacturing (secondary industry)

Nike has factories in 40 countries around the world. Clothing is mainly made in the Asia Pacific area and footwear in China, Indonesia, Vietnam and Thailand. (Just 1 per cent of footwear is made in Italy and no clothing or footwear is made in the USA.) The PT Kukdong International factory in Jawa Barat, Indonesia, is an example of a typical secondary industry in a developing country. The factory makes sports clothes, footwear and equipment for Nike.

Nike has located its manufacturing here for the following reasons:

- Improvements in technology mean that production can be far from Nike's headquarters in Oregon, USA.
- Nike can pay lower wages to workers in developing countries.
- The workforce is more flexible than in developed countries.
- The factory is already there so Nike can subcontract it and not have to build a factory.
- Improvements in transport mean that goods can be manufactured far from the market and then flown or taken by ship to the market.
- Access can be gained to markets all over the world.
- Trade restrictions can be avoided.

How has Nike's presence affected Indonesia?

Benefits to area	Problems for area
Provides jobs	Can cause environmental pollution
Attracts other factories to set up	Influences the decisions of the Indonesian government
Increases Indonesia's wealth	Low wages paid to workers
Provides expert managers	Encourages poor working conditions
Increases exports	Indonesia less inclined to develop its own industries
Increases skill of Indonesian workforce	Workers often sacked without any notice
Improves Indonesia's roads	Sweatshops can develop
Uses latest technology	
May provide healthcare benefits for workers	

How has the Nike's use of labour in developing countries affected more developed countries?

Benefits	Problems
Greater profit for Nike due to low labour costs	Loss of manufacturing jobs in more developed countries
Consumers get cheaper products and greater choice	
Spreads the influence of more developed countries	

Raw materials used in the factory

The factory uses the following raw materials:

- cloth for clothes
- leather for shoes and balls
- thread for sewing
- metal for zips
- buttons.

The factory system for PT Kukdong International Nike factory

Inputs are the things that are needed to make the factory system work, the processes are the actions that take place and the outputs are what is achieved at the end. The linkages are what can be reused by the factory from the outputs.

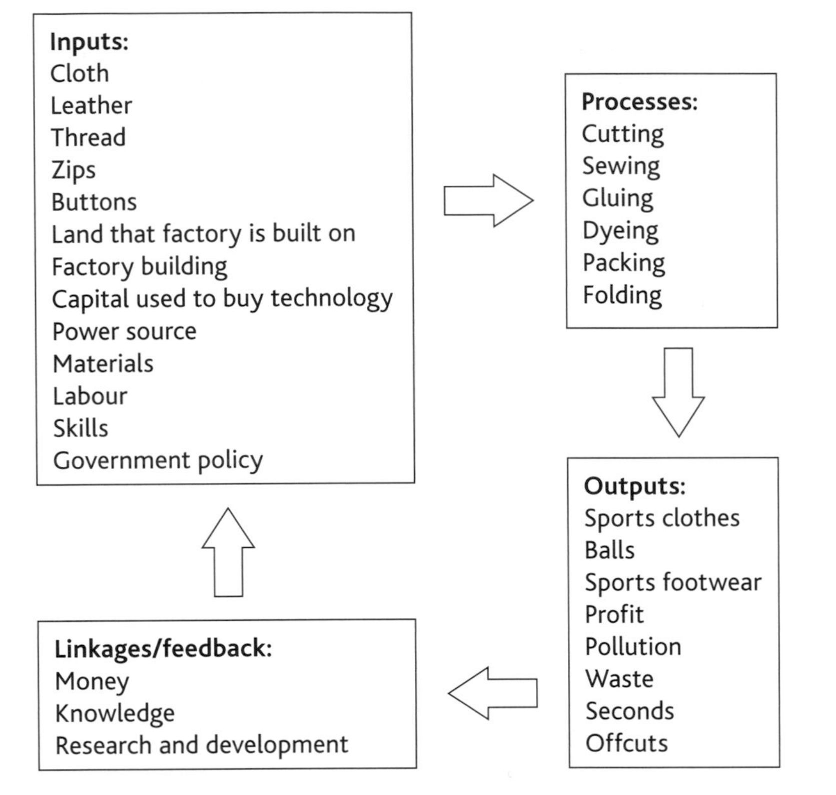

Revision tip

Whichever case study you have chosen it is a good idea to make a mind map of the details. Put the name of the case study in the middle and add a stick for facts, reasons for choices of location, effects on local area, inputs, processes and outputs. Remember to use colour to good effect and to use revision pictures and words rather than sentences.

Exam-style questions

Try these questions. Answers are near the back of the book.

5.1 Why are fewer people involved in primary industry in Britain today than in the last century? (2)

5.2 What evidence of tertiary industry could you find on an OS map? (3)

5.3 Describe the inputs, processes (throughputs) and outputs of an industry you have studied. (4)

5.4 Why is a transport network important? (3)

★ Make sure you know

- ★ The importance of transport and the advantages and disadvantages of different types of transport
- ★ The advantages and disadvantages of a planned or completed transport project
- ★ How economic structure varies according to the level of development of a country
- ★ Factors of location for an industry
- ★ How economic activities may affect the environment and how they can be made sustainable
- ★ An example of a multinational company or an industry in a more developed country and one in a less developed country

Test yourself

Before moving on to the next chapter, make sure you can answer the following questions. Answers are near the back of the book.

1 What is tertiary industry?

2 What is an output?

3 What are the inputs for an industry you have studied in a developing country?

4 Write the definitions of these words and phrases you need to know, and then ask someone to check them. (Words and phrases in italics are useful even though you are not required to know them.)

biodiversity

BRIC countries (Brazil, Russia, India and China)

conserve

containerisation

developed country

developing country

ecosystem

eco-tourism

environment

exploit

globalisation

habitat

HS2 (High Speed 2)

landfill

manufacturing industry

market

multinational corporation (MNC)

national park

newly industrialised country (NIC)

pollution

primary industry

quaternary industry

raw material

recycling

renewable energy

routeway

science park

secondary industry

service industry

stewardship

sunrise industry	tertiary industry
____________________	____________________
____________________	____________________
sunset industry	tourism
____________________	____________________
____________________	____________________
sustainable	*transnational corporation (TNC)*
____________________	____________________
____________________	____________________

6 Ordnance Survey mapwork

6.1 Direction

You need to make sure that you are competent in all of the Ordnance Survey skills.

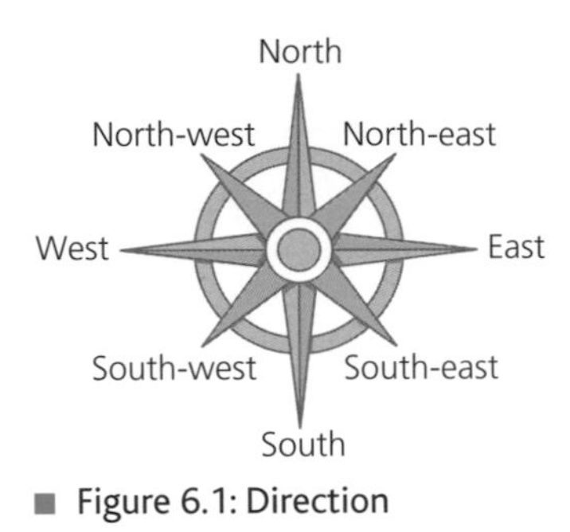

Figure 6.1: Direction

The exam will ask you to state the direction that one feature lies from another. The most important word to note in the question is *'from'*. The easiest mistake to make is to travel the wrong way between the two places.

For example:

- What direction is the church from the town hall?
- Imagine that you are walking *from* the town hall *to* the church and not the other way around!

Revision tip

Make a mnemonic to remember the directions of the compass going clockwise, for example: Naughty Elephants Squirt Water or Never Eat Shredded Wheat!

6.2 Grid references

The exam paper will either ask you to give the grid reference of a feature on the map or it will ask you to find the feature at a particular grid reference.

If the symbol has a stick or arrow attached to it, you must take the reference from the tip of the stick or arrow.

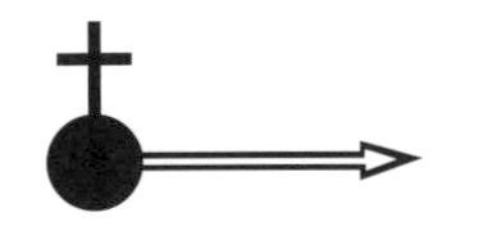

Figure 6.2: Feature with stick or arrow

You may be asked for a four-figure grid reference or a six-figure grid reference.

- A four-figure grid reference refers to a whole square.
- A six-figure grid reference refers to an exact point within the square.

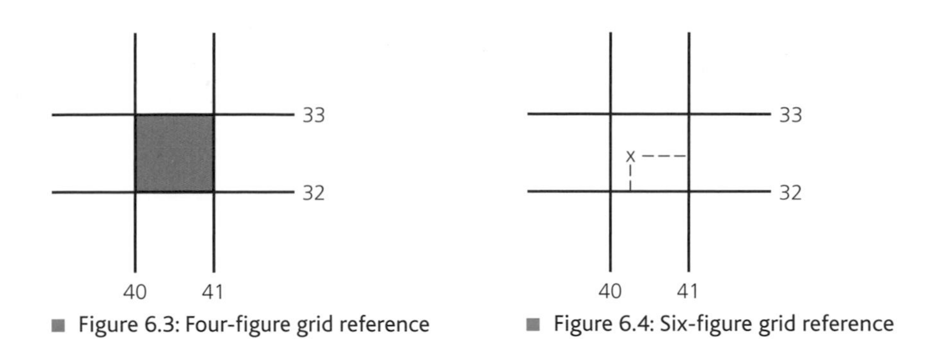

■ Figure 6.3: Four-figure grid reference

■ Figure 6.4: Six-figure grid reference

Always remember to look along the eastings first (along the bottom or top) and then up and down the northings (sides).

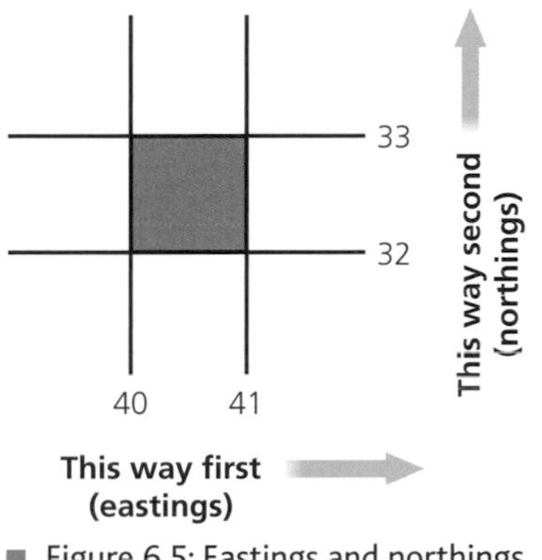

■ Figure 6.5: Eastings and northings

Revision tip

The phrase 'along the corridor and up the stairs' may help you remember the order of eastings and northings when working out a grid reference.

6.3 Distance

Calculating distance is a fairly hard skill to master. If the question asks you for a straight line distance, you can use a ruler to measure in centimetres.

- You then need to put your ruler on the scale of the map and work out the real distance in kilometres.
- Alternatively you can work this out mathematically, converting the centimetres to kilometres.

However, usually you are asked to measure a wiggly line distance along a road or railway. To do this you will need a strip of paper with a straight edge or a piece of string.

- Put a 'start' mark on your strip of paper.
- Position the start mark on the starting point. (Double check the grid reference that you are given for the start.)
- Line up the straight edge of your piece of paper with the first straight section of road, put your pencil point on the end of this section, then work along the road, twisting the paper around your pencil point.

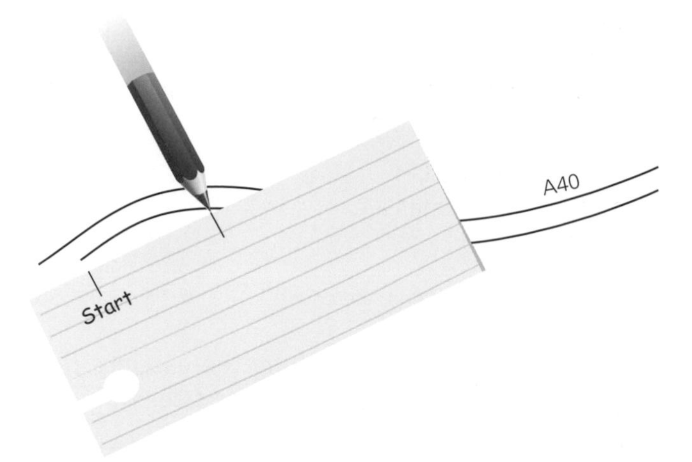

■ Figure 6.6: Measuring a wiggly line distance

- At the 'finish' grid reference, make a mark on the paper.
- Line your piece of paper up on the scale of the map. Ensure that you start at 0 km on the scale.
- If the question asks you to calculate the distance to the nearest whole kilometre, make sure that you do this. If the question does not ask for this, calculate the distance to one decimal place.
- Remember to add the units (km) to your answer.

6.4 Area

In the exam you may be asked to calculate the area of a large feature such as a wood or lake. You are usually given choices and have to tick the box that you think is the most likely area of the feature.

- Look carefully at how many grid squares the feature takes up.
- Each grid square has an area of 1 km².
- Imagine adding part squares together to make whole squares.

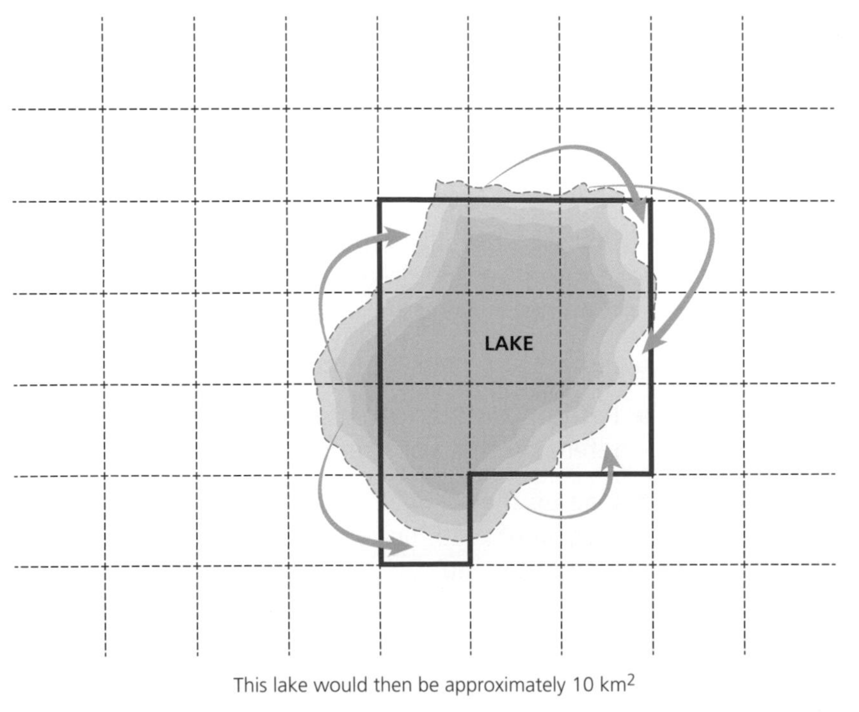

LAKE
This lake would then be approximately 10 km²

■ Figure 6.7: Calculating area

6.5 Height and relief

You may be asked to state the altitude at a certain point or the difference in altitude between two points. There are clues on the map to help you:

- Contour lines are brown lines on the map that join all places of equal height. Not all contour lines are labelled with a height, so you have to calculate their height by looking at the labelled contour lines or spot heights around them.
- Spot heights are black numbers on the map that indicate the exact height at a certain spot.

If you are asked to work out how high you have climbed if you walk from one location to another, you need to subtract the starting point height from the finishing point height.

Remember to add the units (m) to your answer.

If you are asked to describe or compare the relief in two parts of the map then you need to imagine the landscape that the contour lines are creating in 3D. Use words such as flat, undulating, hilly, mountainous, valley, plateau and ridge.

6.6 Cross-sections

You may be asked to draw something such as a road, path or woodland onto a cross-section. To do this you will need a piece of paper with a straight edge.

- Line up the straight edge of your piece of paper along the bottom axis on the cross-section that you have been given.
- Add a start mark and a finish mark on your piece of paper. Write down the grid references for the start and finish points.
- Place your piece of paper between these two grid references on the map.
- Mark onto the piece of paper where the road, path or woodland touches the paper.
- Mark on the height of the land as you move along the piece of paper.
- Then position the piece of paper back onto the cross-section and mark with an arrow or a bracket the correct location of each feature. Annotate the cross-section with the names of the features.
- Mark the height of the land on to the cross-section and join up the dots to show the relief of the section covered.

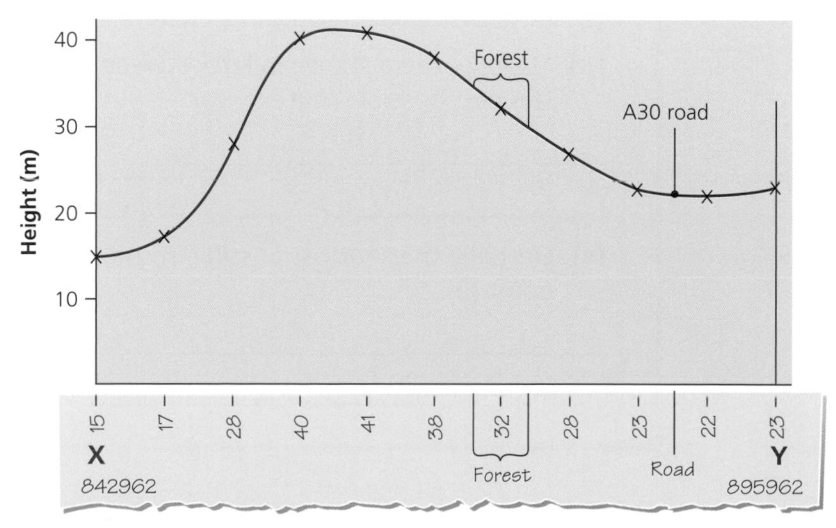

Figure 6.8: Drawing a cross-section

You may also be asked to annotate simple sketch sections.

★ Make sure you know

★ How to identify features at four- and six-figure grid references

★ How to give four- and six-figure grid references for features

★ How to calculate distance

★ How to calculate area

★ How to give a direction from one place to another

★ How to estimate altitude

★ How to describe the relief of an area

★ How to annotate cross-sections and sketch sections

Test yourself

Before moving on to the next chapter, make sure you can answer the following questions. You will need to use an Ordnance Survey map (1:50 000 or 1:25 000).

1 (a) If the map is of your local area, find the location of either your house or your school, and write the six-figure grid reference number. If the map is of a different area, choose any building or landmark.

____

____

(b) Find an example of an industry on the map, and write the six-figure grid reference number.

____

____

2 Using the two points marked in questions 1 (a) and (b):

(a) Measure the distance in km between the two points as the crow flies.

____

____

(b) Measure the distance in km between the two points of the shortest possible route by car.

____

____

(c) Describe the shortest possible route by foot from point (a) to point (b).

____

____

3 Using the two points marked in questions 1 (a) and (b):

(a) What is the direction of point (a) from point (b)?

____

____

4 Write the definitions of these words and phrases you need to know, and then ask someone to check them.

compass

____

____

northing

____

____

contour line

____

____

OS (Ordnance Survey)

____

____

easting

____

____

relief

____

____

grid reference

____

____

symbol

____

____

key

____

____

Revision tip

You could play Pictionary with your friends to help you learn the words you need to know. One leader writes ten words in a list and then one member of each team comes to the leader and is given one word to 'draw' to their team. Once the correct word has been guessed another member of the team comes to collect another word to 'draw'. The first team to finish all ten words wins. No letters or numbers are to be drawn and the scribe cannot speak!

7 Fieldwork

7.1 Fieldwork investigation

As part of the process of learning geographical skills, you will complete a fieldwork investigation. This accounts for 20 per cent of the 13+ Common Entrance examination mark. It enables you to show those geographical skills that cannot be examined in the written papers – especially graphical skills. It is also a great deal of fun!

Your geography teacher will probably arrange some time away from the normal school day to collect the data for your project. You may go to a field centre some distance from your school, you may go abroad or you may collect the data in your local area.

You should submit your project as bound A4 sheets or saved onto a CD-ROM.

When do I complete the investigation?

You will probably be given a number of weeks in which to write up your fieldwork investigation after you have collected the data. It is best to get the majority, if not all, of the investigation done while the data collection stage is fresh in your mind. You will be given a deadline for the whole investigation to be complete. This may be broken down into deadlines for each of the sections in the project.

If you are sitting Common Entrance in February, the deadline for submission of your project to your senior school is January; if you are sitting Common Entrance in the summer, then the deadline is March.

How do I plan my investigation?

You will probably plan your investigation with your class and teacher, and will probably all set out with the same aims or question to be answered.

Your question to be answered may be in the format of a hypothesis – a statement that you are going to 'test' and will either accept or reject at the end of your project.

The suggested word limit for your project is 1000 words. Part of the skill of writing a fieldwork investigation is being concise.

Your investigation should be divided into distinct sections, and is marked as such.

Section 1 Introduction

You should include:

- a clearly stated question to be answered or a hypothesis
- a reason why you think this is a suitable question to be asked
- a brief description of the area of investigation
- information on the topic
- your geographical aims
- a prediction of the outcome based on present knowledge.

Section 2 Methods of data collection
You should include the following:

- A map of the area. A hand drawn map is a good idea. If you use one from a computer make sure you annotate it. All maps should include a title, key, scale and north arrow.
- An explanation as to why this area was chosen as a suitable area for carrying out the investigation.
- An explanation of the methods that you used for collecting your data, and a justification for these methods.
- A photo of the methods used, with annotations.

Section 3 Results/presentation of data
You should include at least two different ways of presenting your data.

- Appropriate and accurate charts, graphs, cross-sections and tables can be used.
- Marks will be awarded for innovative ways of presenting your data.
- Everything must have a heading and labelling.

Section 4 Data analysis and conclusion

- For each graph, chart or table explain any patterns that emerge and any odd results (anomalies), with suggested reasons for why they occurred.
- Answer the question or confirm or reject the hypothesis that you stated at the start of the project. Give the geographical reasons for this.
- Explain the limitations of your project. Consider how you could improve what you did. Consider what might change your outcomes.
- Give references for any resources that you have used, including books, maps, software programs and secondary resources.

7.2 Fieldwork tips

- Listen carefully to instructions and information from trip leaders and teachers.
- Ask questions on your fieldtrip, and be alert to your surroundings at all times.
- Make concise notes on your return to the field centre or your classroom, while the trip is still fresh in your mind.
- Neatness, punctuation, spelling and general presentation are all taken into account when it comes to marking.
- All work should be presented on A4 paper. Smaller or larger formats are not allowed. All maps, charts, etc. must also be presented on A4. Do not include folding maps – it is better to photocopy the relevant area and insert it on A4 paper.
- Draw some of your graphs, pie charts, tables and sketches by hand to demonstrate that you possess that skill. Contrast this by using the computer. The examiner is looking for a variety of presentation techniques.
- Use suitable coloured pencils when shading.
- Do not forget appropriate titles, keys, scales, direction and labelling.

- Always save your information on a memory stick and a hard drive.
- If in doubt about any aspect of your fieldwork investigation, ask your teacher for clarification. Remember, however, that it is your work that will be marked.
- Manage your time effectively and remember that deadlines are deadlines. A rushed job can be spotted a mile off.
- Be enthusiastic and put your skills, time and effort into your investigation and you will produce a piece of work you can be proud of.

7.3 Marking of fieldwork

The investigation is moderated by your geography teachers and given a mark out of 20. A breakdown is shown below:

Introduction	4
Data collection	8
Data presentation	8
Data analysis	12
Fieldwork expertise	8
Total mark	**40**

This total mark (out of 40) is then halved to give a mark out of 20 for the investigation. This mark is added to your written exam mark (which is out of 80) to give a percentage.

This process takes a long time, as each individual investigation has to be thoroughly checked. Investigations and assessment forms are sent to senior schools. On the assessment form, your teacher is asked to include written information about the amount of assistance you have been given. Your teacher will be accurate and truthful when it comes to moderating and answering this question.

Test yourself ✓

Write the definitions of these words and phrases you need to know, and then ask someone to check them. (Words and phrases in italics are useful even though you are not required to know them.)

analysis

____

____

methods

____

____

fieldwork

____

____

percentage

____

____

graph

____

____

primary data

____

____

justify

____

____

secondary data

____

____

Exam-style question answers

Chapter 1

1.1 See Figure 1.9. (1)

1.2 *Any four from*:

- North and South American plates met the Caribbean plate and were subducted underneath it.
- This is a destructive plate boundary.
- As the oceanic plates subduct the friction causes the plates to melt.
- This causes excess magma which rises as it is full of gas bubbles which make it lighter than the surrounding rock.
- The magma forces its way to the surface forming the Soufrière Hills volcano. (4)

This is better answered using a well-annotated (labelled) diagram like Figure 1.10.

1.3 Soufrière Hills volcano on Montserrat

Any four from:

- Pyroclastic flow burned vegetation.
- Ash covered two-thirds of the island, destroying agricultural land.
- Coral reef and sea creatures died from the ash washed into the sea.
- 60 per cent of housing was destroyed.
- Hospitals/schools were destroyed. (4)

1.4 A Mid-Atlantic Ridge

B Pacific Ring of Fire (2)

1.5 *Any three from:*

- Lack of quick reaction forces to rescue people.
- Poor medical care and hospitals.
- Population is densely packed around a volcanic cone to benefit from the fertility of the soil.
- Lack of technology and money for prediction equipment. (3)

Chapter 2

2.1 A humid tropical is much hotter and wetter (convectional rainfall) and does not have a seasonal pattern of temperature. (3)

2.2 Places with humid tropical climates are near the Equator, therefore the Sun's rays are more concentrated and they are hotter.

Diagram could be included.

As humid tropical places are hotter there is rapid evaporation, which leads to convectional rainfall.

Humid tropical places are on the Equator and therefore they are never tilted away from or towards the Sun and do not have seasonal temperatures. (3)

2.3 (c) microclimate. (1)

2.4 *Any three from:* aspect; proximity to buildings; surfaces; distance from sea; whether in an urban or rural area. (3)

2.5 The local climate might vary during the course of a bright, sunny day, depending on the physical features, surfaces and aspect. If there is a lot of tarmac, this will absorb heat throughout the day and release it at night, making night-time temperatures warm. In rural valleys, cold air will sink at night, causing frosts, and will take time to warm up through the day. A south-facing slope or wall will obviously have the benefit of the Sun throughout the day so temperatures will increase from the start of the day right through to evening. (4)

Chapter 3

3.1 **(a)** Corrosion, abrasion or attrition. (1)

(b) Corrosion occurs when river water (which is slightly acidic) dissolves particles of rock.

Abrasion takes place when small pieces of material rub against the bed and banks of the river.

Attrition occurs when particles collide and knock pieces off each other. (3)

3.2 The material transported in a river is the load. (1)

3.3 As the river channel moves through the river basin, from source to sea, the size and shape of the load alters. To start with stones roll along the bed. These stones are worn down until they become round and smooth and eventually become small particles, which 'leap-frog' along the bed. These particles become smaller and smaller until they form a suspension within the water flow, and eventually are small enough to be dissolved in the water. (3)

3.4
- Waves attack the fault by hydraulic action (sheer force of the waves hitting the headland and forcing air into cracks) and by abrasion (load being carried by waves hitting the headland) and corrosion (acids in sea water attacking the headland).
- Fault enlarged by the same three processes to form a cave.
- Cave is widened and deepened by the same processes to form an arch.
- Undercutting and gravity leads to collapse, leaving a stack.
- Weathering (freeze-thaw, chemical or biological) and erosion turn the stack into a stump. (5)

3.5
- A corner in the coastline, weak currents, shallow water and longshore drift are all required for spits to form.

- The processes involved in spit formation are transportation (longshore drift) and deposition.
- *Diagram required in the answer to show how longshore drift creates spits (see Figure 3.15).* (5)

Chapter 4

4.1 *Any three from:*

- temperate climate
- flat land
- fertile soil
- many natural resources
- lack of natural disasters
- stable government
- many job opportunities (3)

4.2 Answer should be the current world population. (1)

4.3 City (1)

4.4 Linear (1)

Chapter 5

5.1 *Any two from*:

- Mechanisation of farming
- Imports of food from abroad, fewer farmers in UK
- Crises in farming, such as Foot and Mouth
- Decline in coal mining as cheaper, cleaner, more efficient fuel becomes popular (2)

5.2 *Any three from*:

- School
- Hospital
- Recreation/leisure centre
- Tourist information centre
- Any tourist feature
- Golf course (3)

5.3 Answer could refer to the inputs, processes (throughputs) and outputs of the factory system for PT Kukdong International Nike factory (see Case Study 5.10 in Chapter 5) or any other case study. (4)

5.4 *Any three from*:

- People need to get to work (commute) using transport infrastructure and children need to get to school.
- People who live in rural areas need to access urban areas for shops and services.

- People use transport for leisure/tourism and for socialising.
- Industry needs transport for accessing raw materials, for getting the finished product to the market and for labour to get to work.
- Quality of life is improved for people who have good access to transport infrastructure and it also brings an increase in the value of property. (3)

Test yourself answers

Chapter 1

1 The Pacific Ring of Fire

2 Due to convection currents

3 A tiltmeter

Chapter 2

1 (a) Two from:

- latitude
- altitude
- prevailing wind
- distance from sea
- ocean currents

(b) Weather is the hour-to-hour, day-to-day condition of the atmosphere (wind speed, wind direction, temperature, humidity, sunshine, type of precipitation).

Climate is the average weather conditions for an area over a long period of time.

2 (a) Infiltration

(b) Surface run-off

3 (a) Precipitation

(b) South-west

(c) An ocean current

(d) In winter

(e) Relief rainfall

Chapter 3

1 Velocity is the **speed** of the water. Deposition is the 'dumping' of a load when the river's velocity **reduces.** Load is the material which the river **transports.** Load can be deposited by rivers at their mouth; the feature formed is called a **delta**. It is also deposited on the **inside** bend of a meander; the feature formed is called a **river beach** (or **slip-off slope**).

2

	Upper course	Lower course
Size of load	large	fine
Shape of load	angular	smooth
Main methods of transportation	traction	suspension
	saltation	solution

3 (a) Chemical weathering

(b) Freeze-thaw

(c) Onion-skin (exfoliation)

4 River basin – an area of land drained by a river and its tributaries

Watershed – the boundary of the river basin, usually marked by a ridge of high land

Source – the start of a river

Mouth – where a river meets the sea

Permeable – rocks that allow water to pass through

Impermeable – rocks that do not allow water to pass through

Evaporation – the loss of water to the air when the water has turned into water vapour

Transpiration – the loss of moisture to the air from plants

Through flow – the movement of water through the soil back to the sea

Ground water storage – water stored in rocks below the ground

Infiltration – the downwards movement of water through tiny pores in the soil

Surface run-off – the movement of water over the surface of the land back to the sea

River discharge – the amount of water that passes a given point at a given time, measured in cumecs (cubic metres per second)

Load – the material that a river carries

Attrition – when the river's load collides and breaks into smaller pieces

Corrosion – a type of erosion caused by the acids in the river dissolving the rocks

Hydraulic action – a type of erosion caused by the force of the water breaking particles of rock from the river bank

5 (a) Hydraulic action

(b) Spit

(c) Stack

6 See Figure 3.14.

7 Answers will vary according to the case study used.

Chapter 4

1 Initially population was low due to war, famine and disease, even though family size was high. Death rate then declined rapidly in the nineteenth century due to medical advances. This meant that the population increased at a rapid rate as the birth rate was still high. In the early twentieth century, the birth rate began to fall as women began to have careers and infant mortality rates declined reducing the need for many children. Population grew more slowly but this was boosted by immigration from the Commonwealth and Jews. Population is now high

but stable, but immigration is still increasing UK population figures, especially from countries within the EU.

2 In the past, people needed to live in groups as protection from invaders, so a settlement would grow up around a feature such as a spring or a crossroads, and therefore would become nucleated at that point.

3 Higher up the hierarchy a settlement has more services, for example a town has more shops than a village and a city has more schools and places of worship than a town.

Chapter 5

1 A service industry selling goods or providing a service

2 Something that a factory produces, for example, profit, clothes

3 *Answers will vary according to the case study used.*

Appendix 1: Location knowledge maps

The maps on the following pages show the location information you will be required to know for your exam.

Map 1: UK, Great Britain and British Isles

This should be known by the end of Year 6.

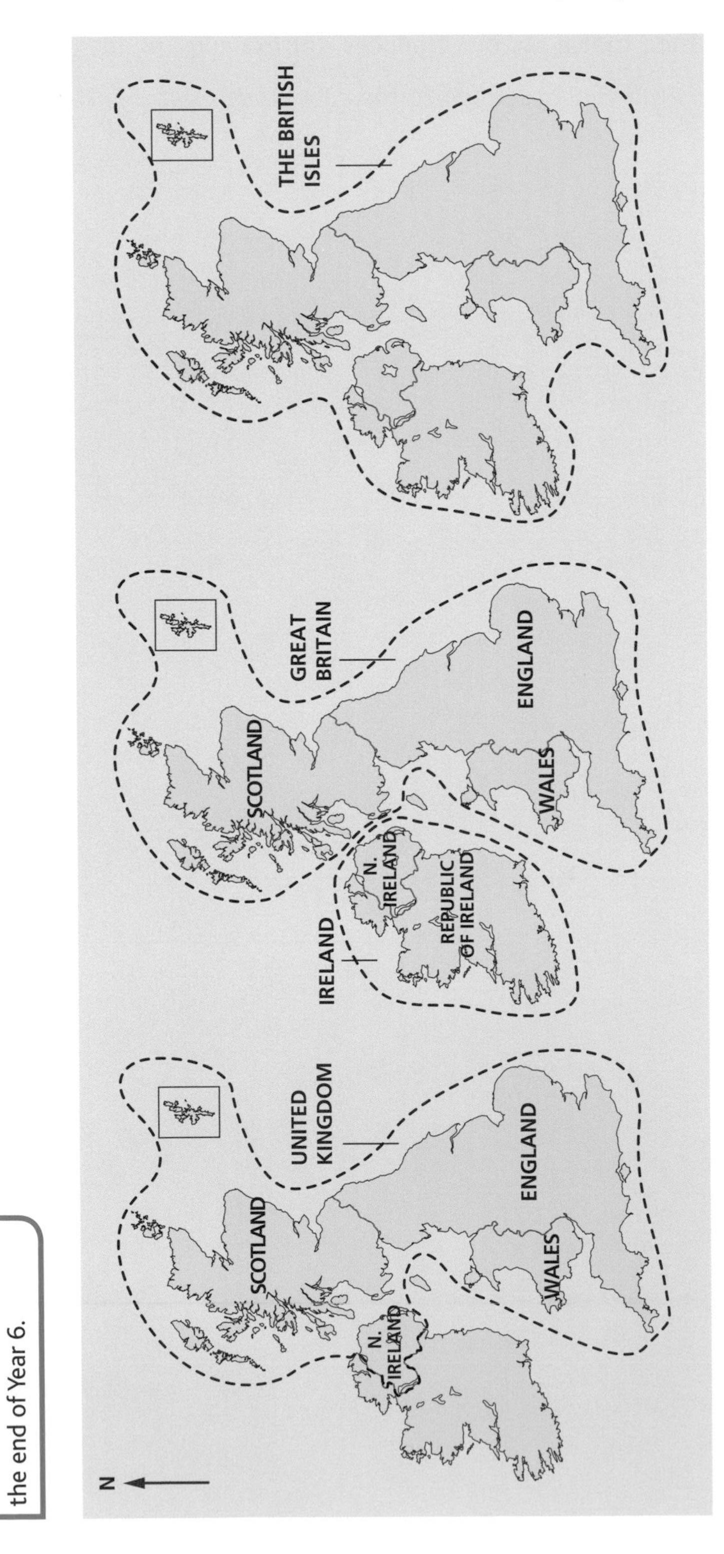

Map 2: Continents and oceans

This should be known by the end of Year 6.

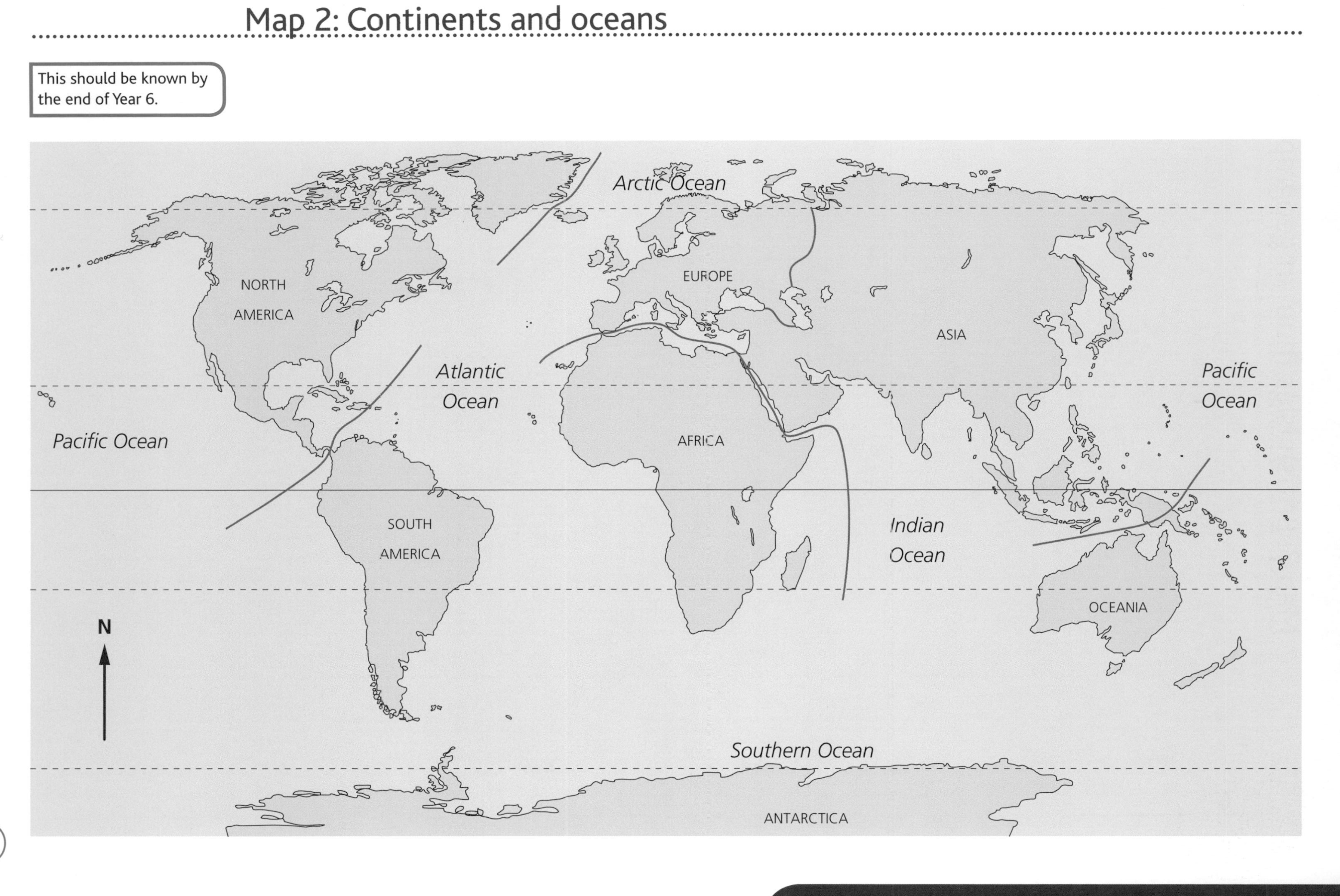

Map 3: British Isles (human geography)

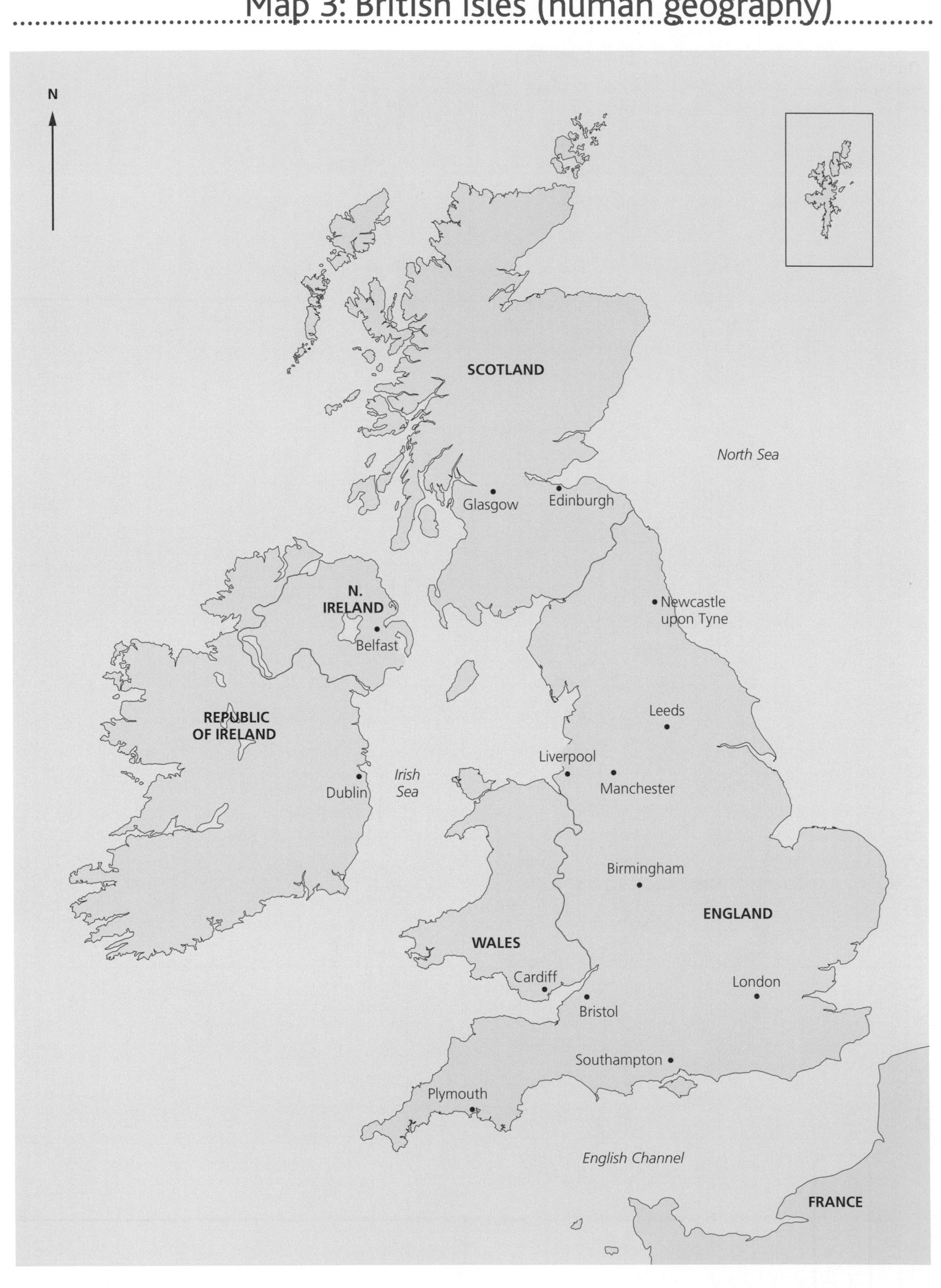

Map 4: British Isles (physical geography)

This should be known by the end of Year 6.

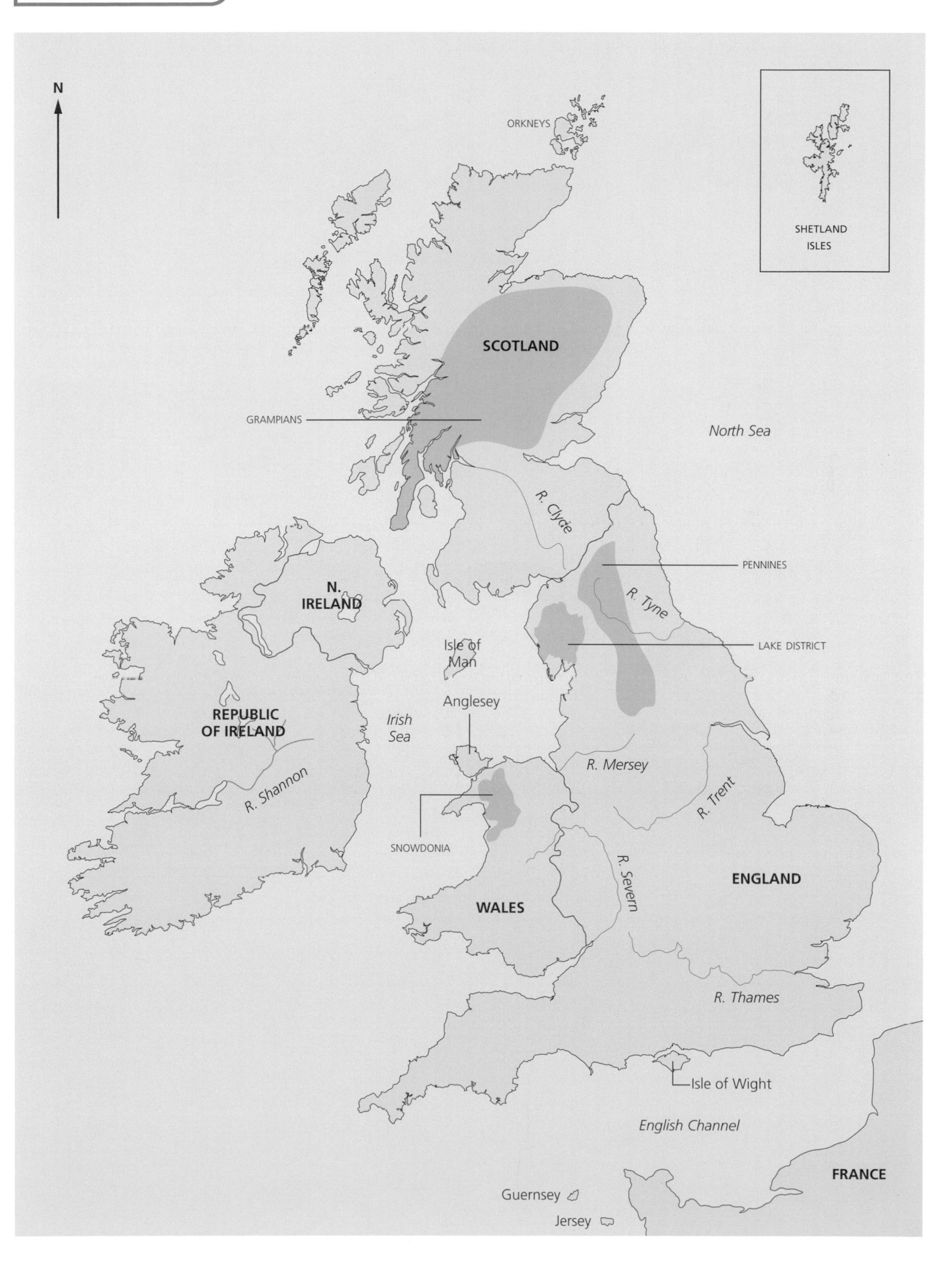

Map 5: Europe (physical geography)

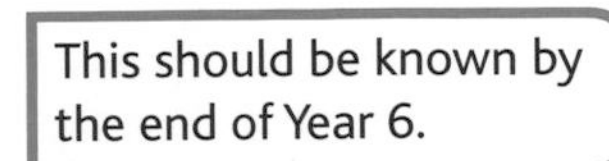

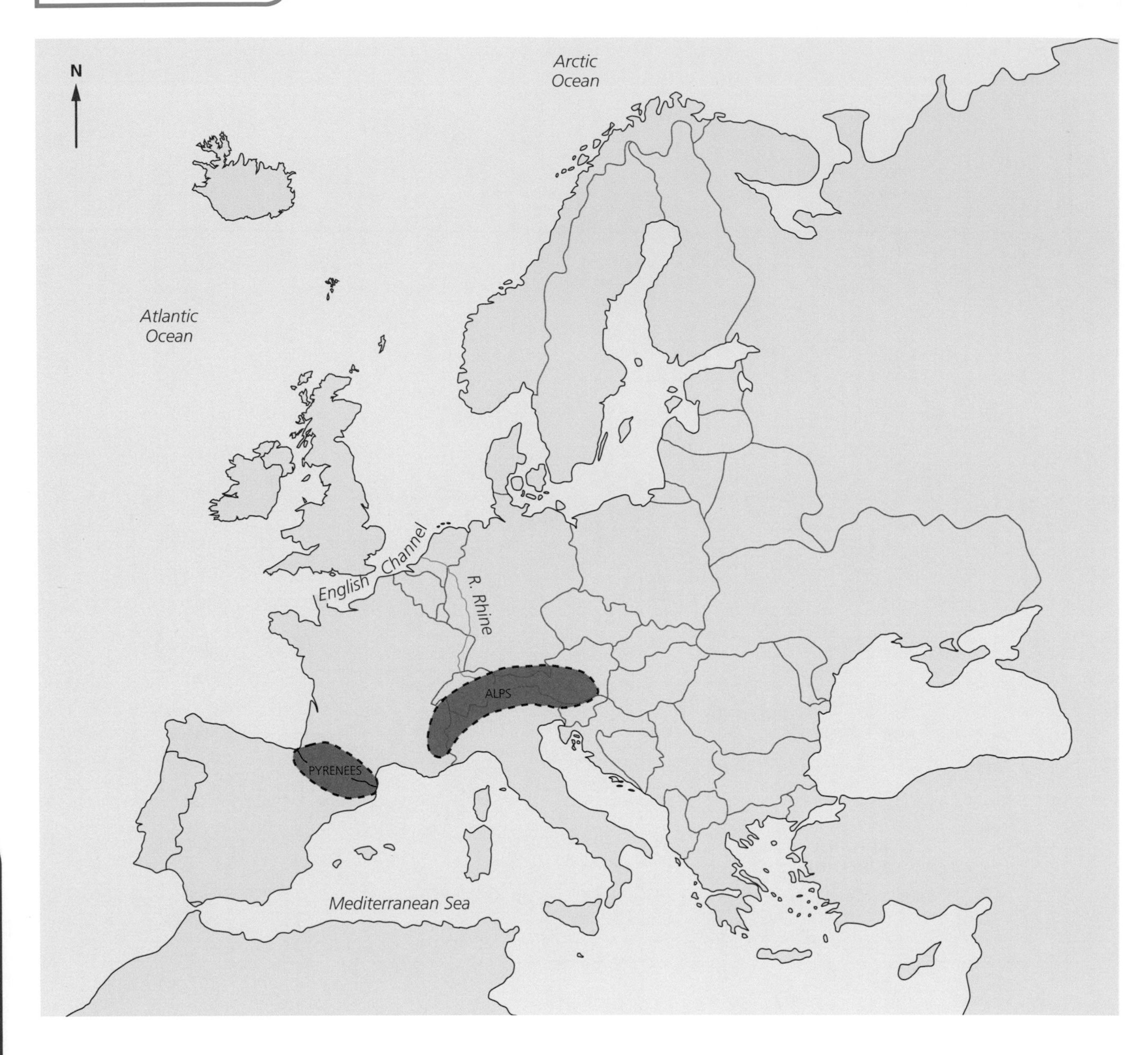

Map 6: Europe (human geography)

This should be known by the end of Year 6.

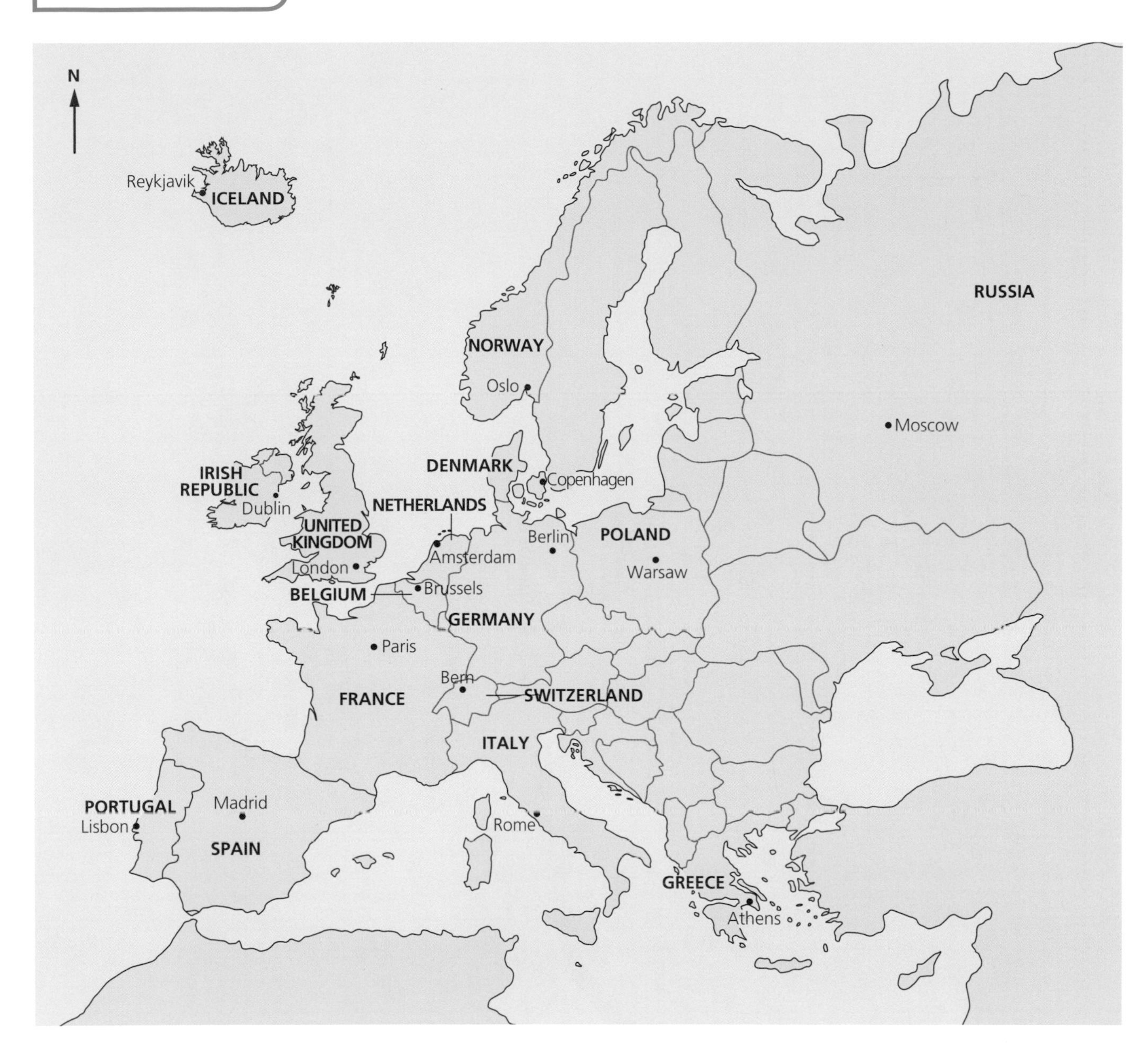

Map 7: Asia

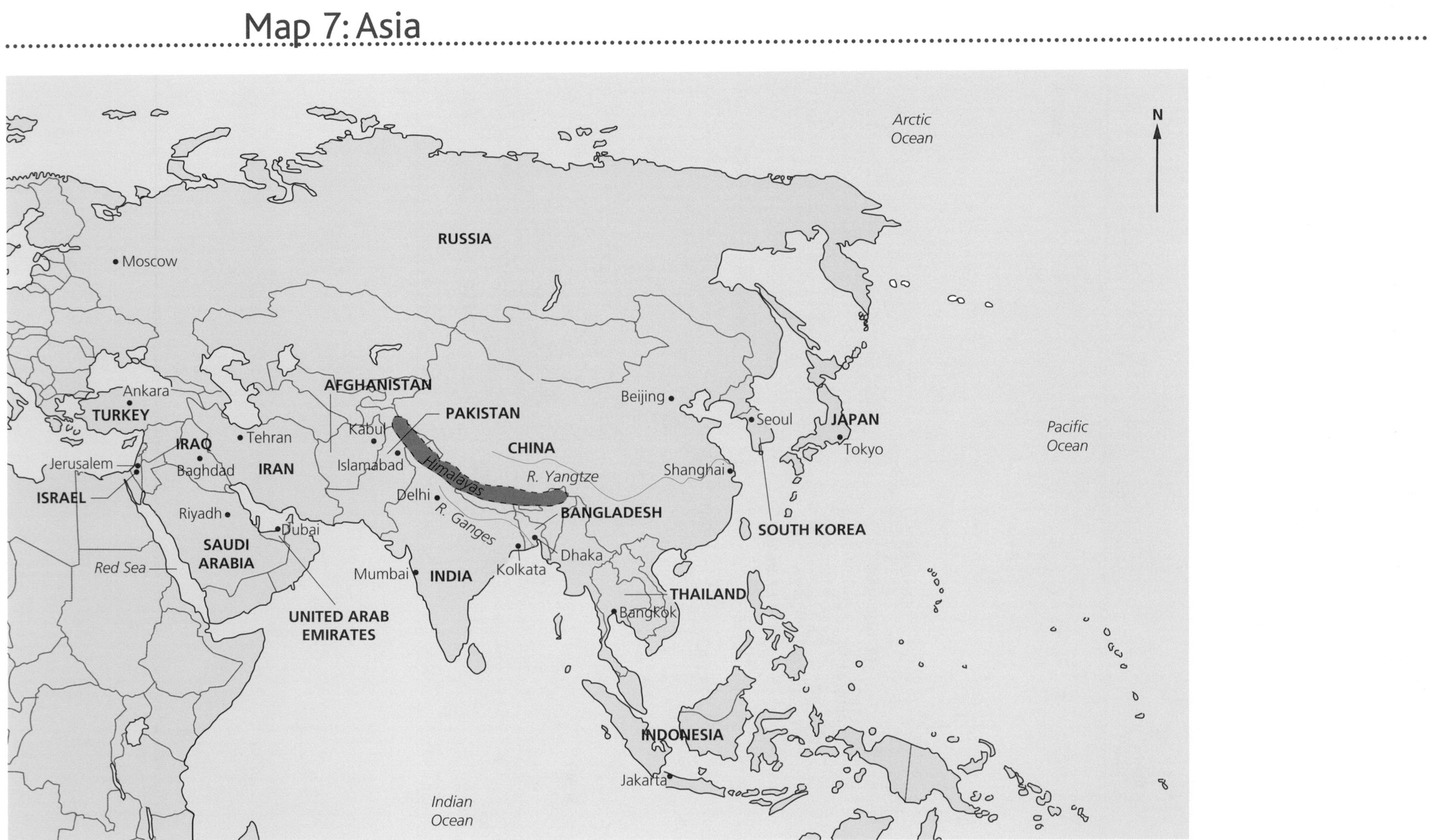

Arctic Ocean
N
RUSSIA
Moscow
AFGHANISTAN
PAKISTAN
Beijing
Ankara
TURKEY
IRAQ
Tehran
Kabul
Seoul
JAPAN
Pacific Ocean
CHINA
Tokyo
Jerusalem
Baghdad
IRAN
Islamabad
Himalayas
R. Yangtze
Shanghai
ISRAEL
Delhi
Riyadh
R. Ganges
BANGLADESH
SOUTH KOREA
Dubai
SAUDI ARABIA
Dhaka
Red Sea
Mumbai
INDIA
Kolkata
THAILAND
Bangkok
UNITED ARAB EMIRATES
INDONESIA
Jakarta
Indian Ocean

Map 8: Oceania

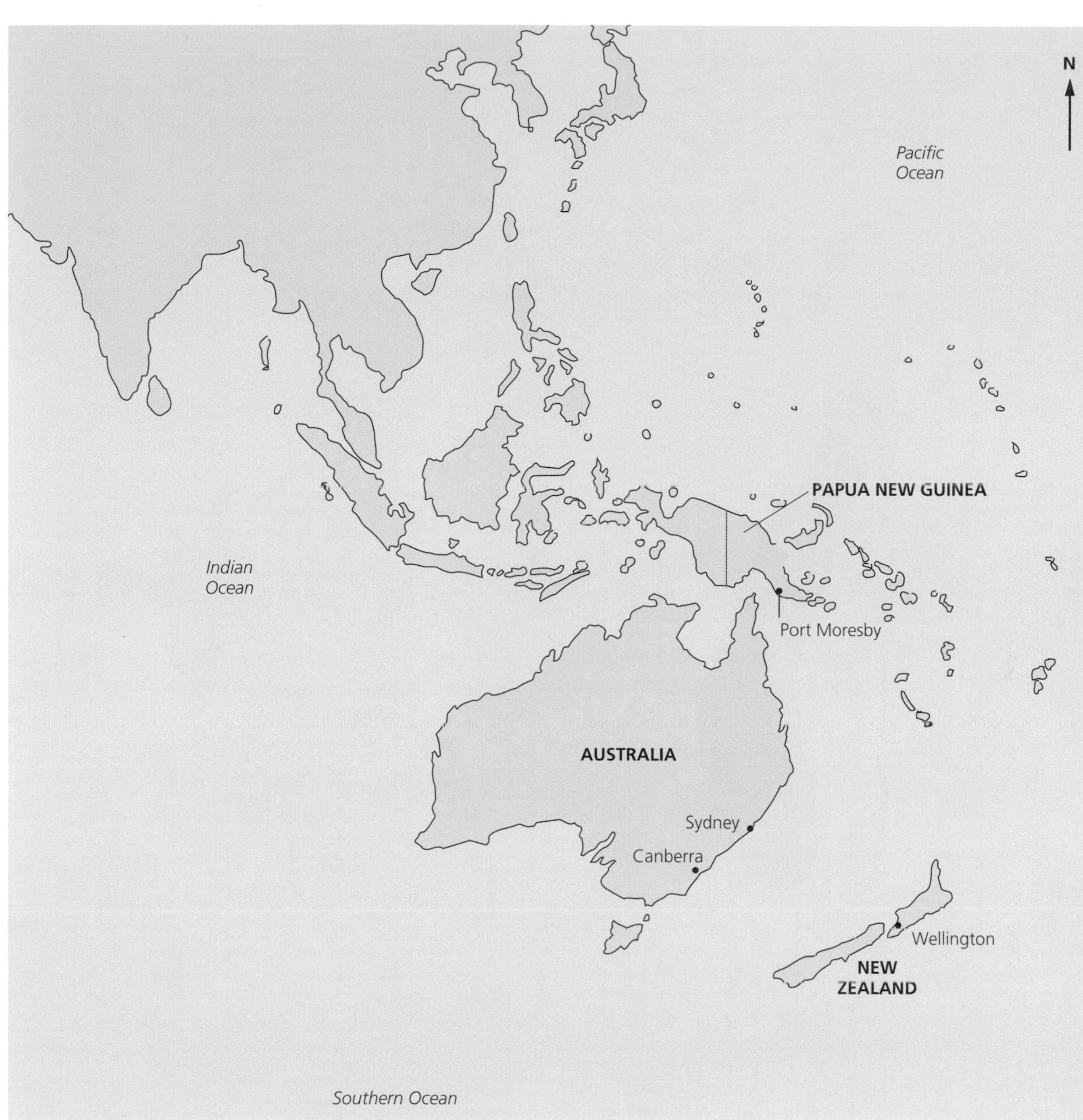
N
Pacific Ocean
PAPUA NEW GUINEA
Port Moresby
Indian Ocean
AUSTRALIA
Sydney
Canberra
Wellington
NEW ZEALAND
Southern Ocean

Map 9: North and Central America

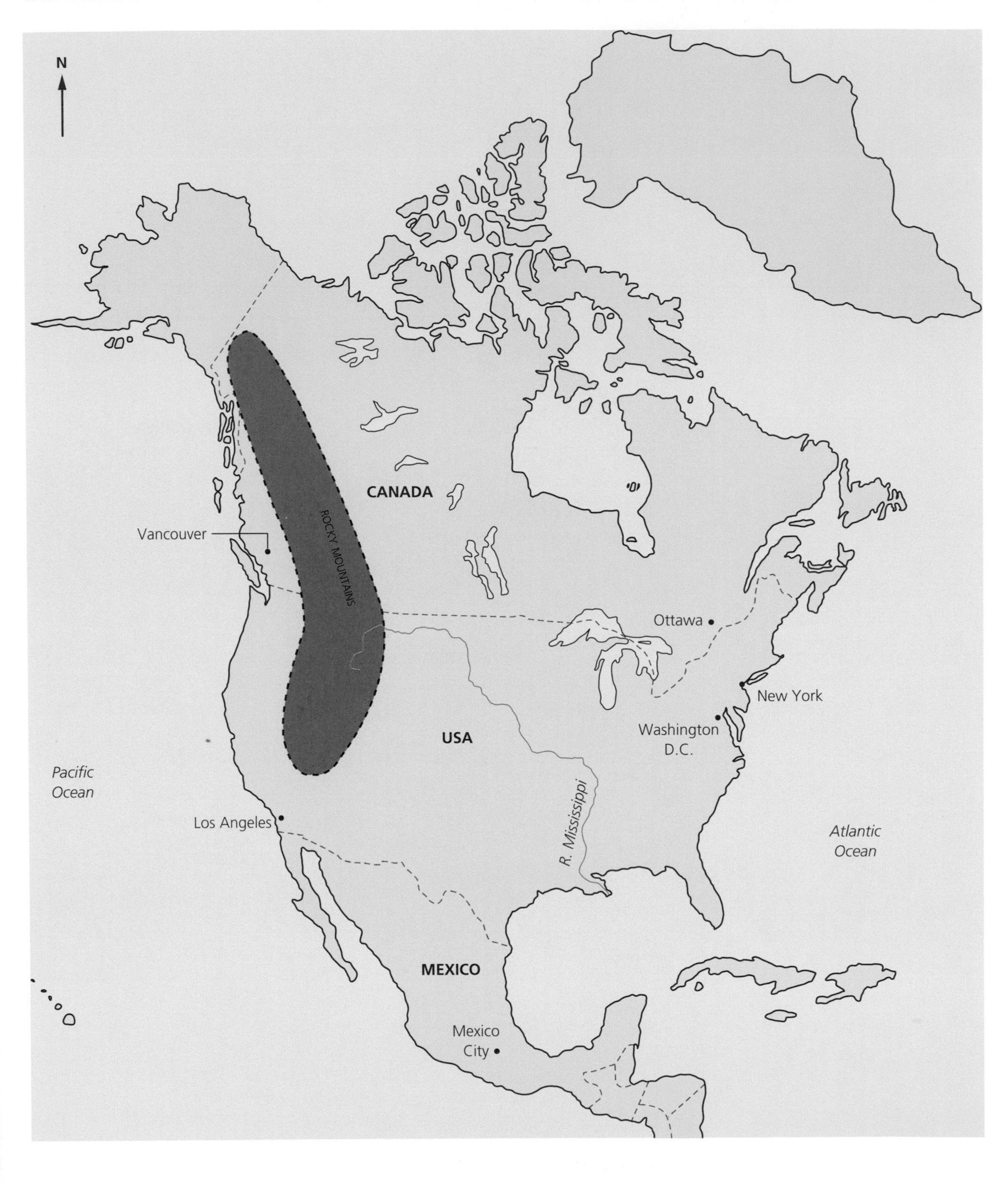

Map 10: South America

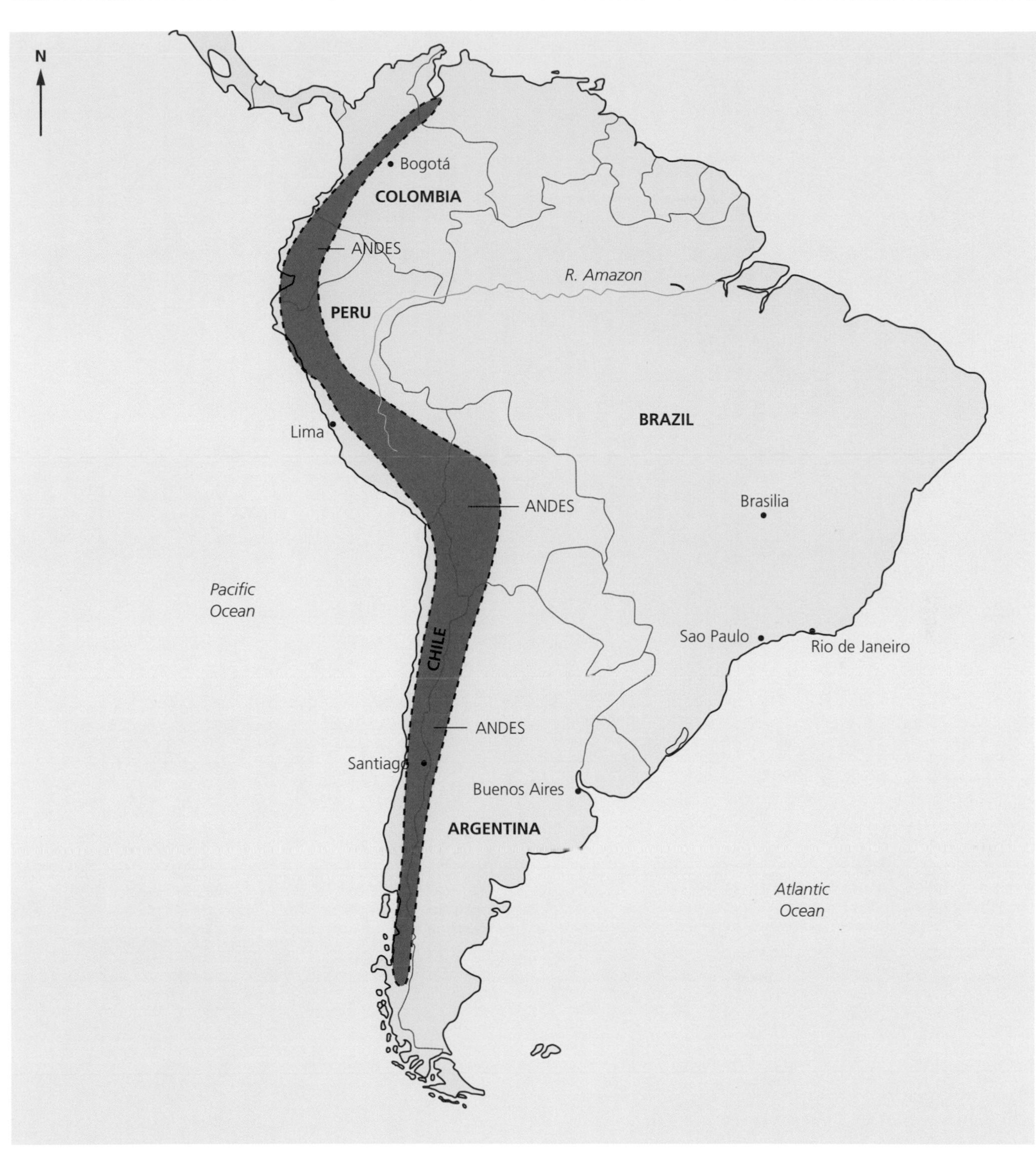

Map 11: Africa

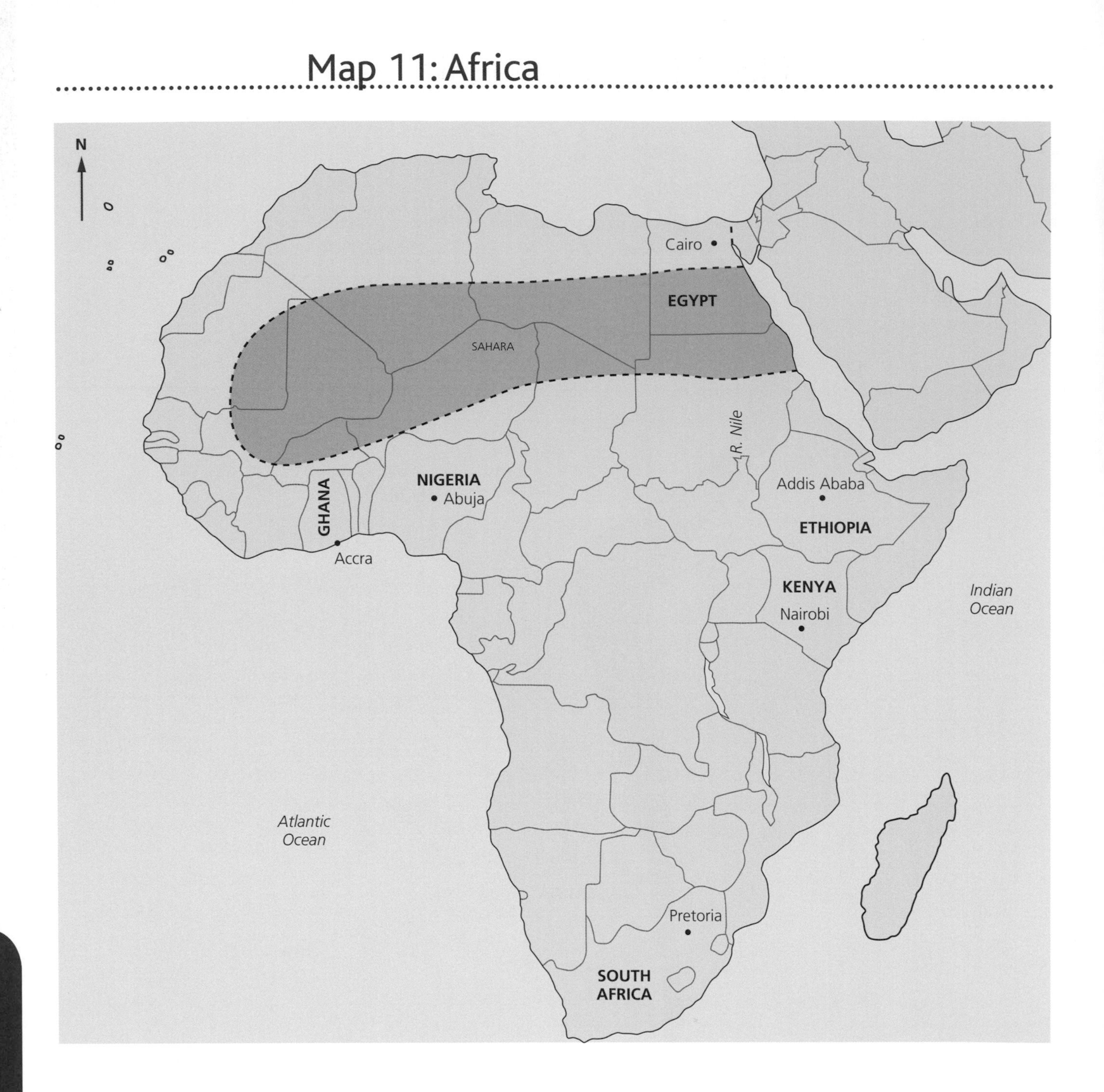

Map 12: General world features

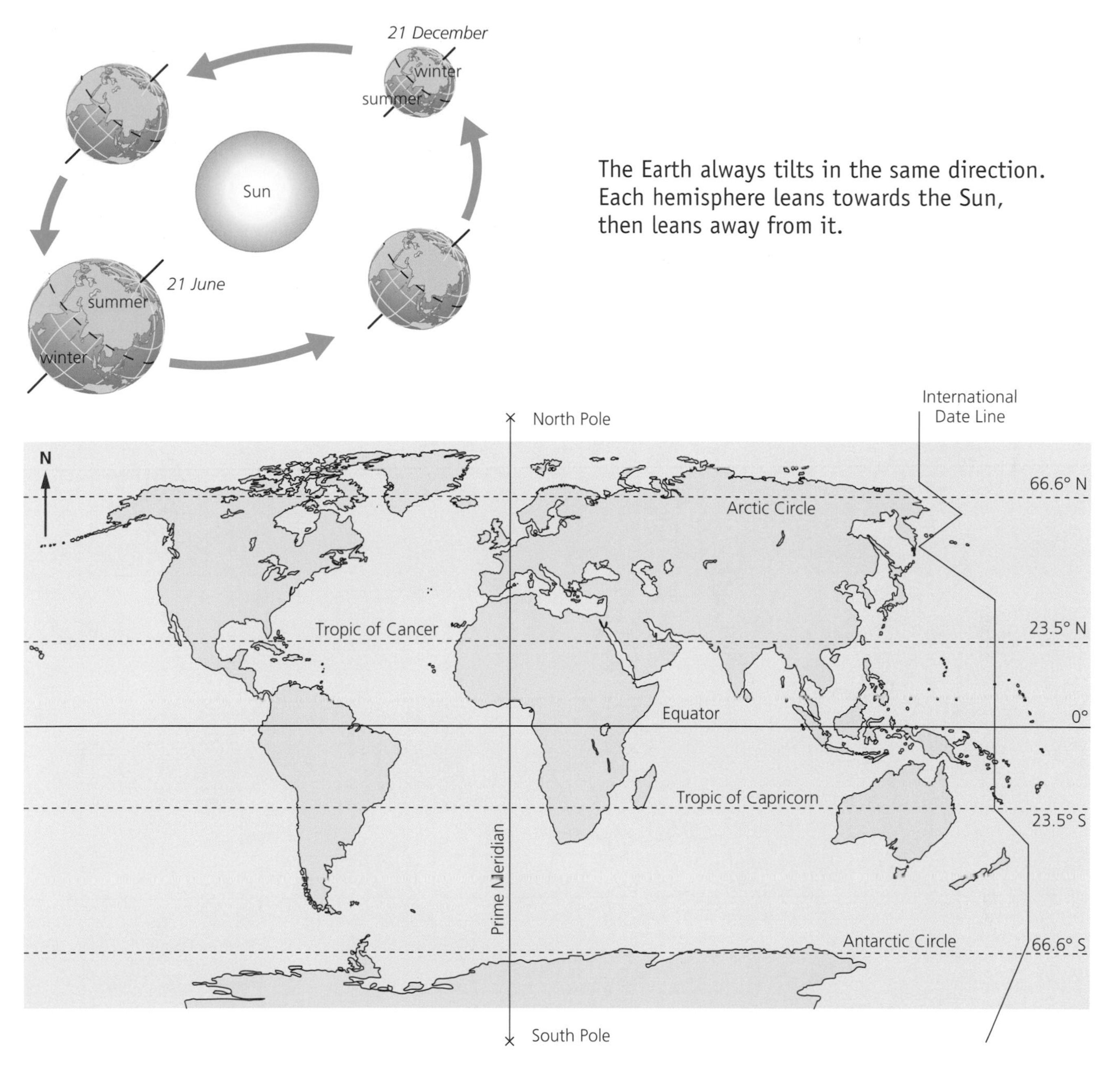

The Earth always tilts in the same direction. Each hemisphere leans towards the Sun, then leans away from it.

Appendix 2: Blank maps

The maps on the following pages are blank copies of the maps in Appendix 1 for you to practise with.

Map 1: British Isles

Map 2: World map

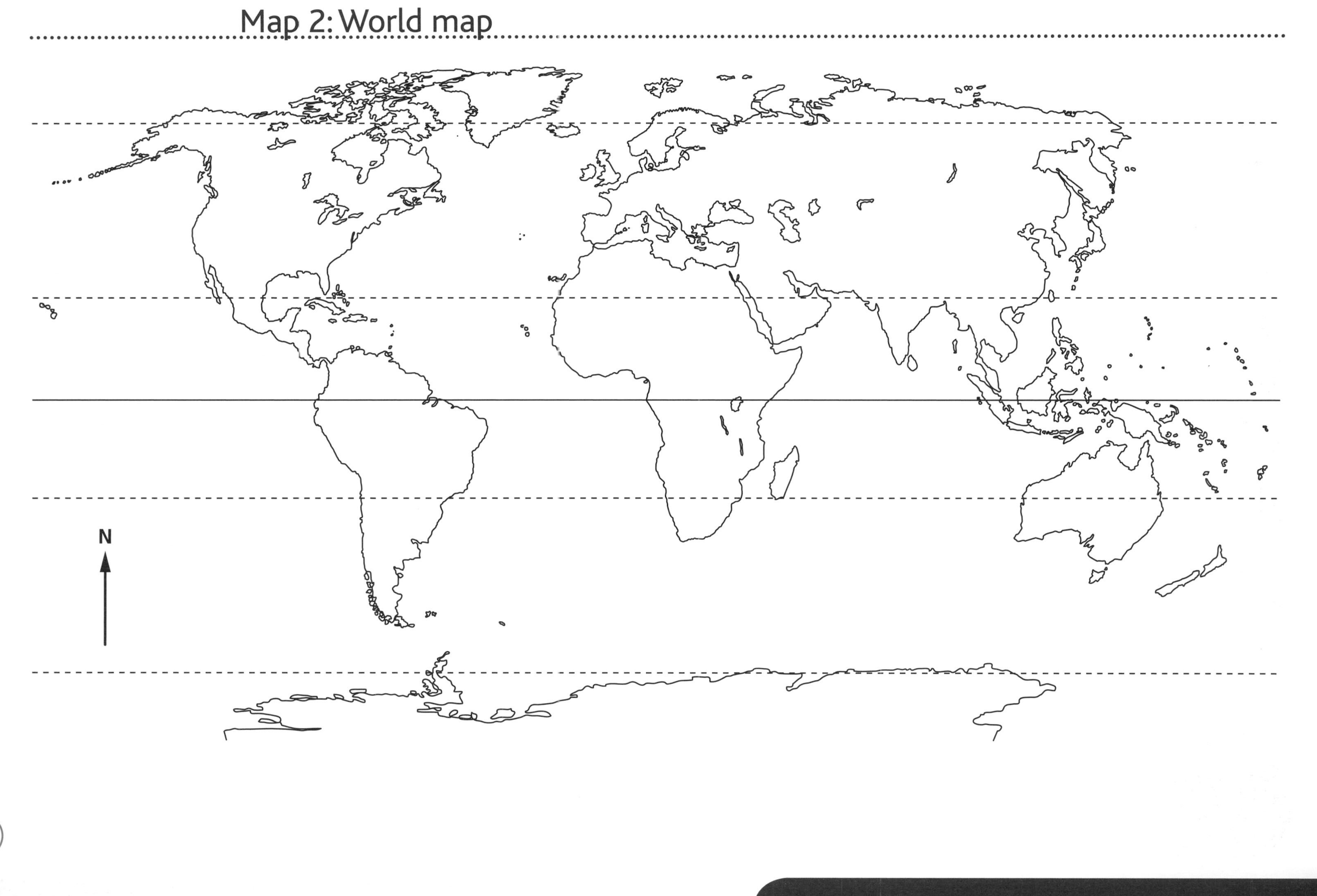

Map 3: British Isles (human geography)

ENGLAND
SCOTLAND
WALES
REPUBLIC OF IRELAND
N. IRELAND

Belfast
Birmingham
Bristol
Cardiff
Dublin

Edinburgh
Glasgow
Leeds
Liverpool
London

Manchester
Newcastle upon tyne
Plymouth
Southampton

Map 4: British Isles (physical geography)

English Channel
Irish Sea
North Sea

R. Clyde
R. Severn
R. Shannon
R. Thames
R. Trent
R. Tyne
R. Mersey

Grampians
Lake district
Pennines
Snowdonia

Guernsey
Isle of Man
Isle of Wight
Jersey
Orkneys
Shetland Isles

Map 5: Europe (physical geography)

R. Rhine

Alps
Pyrenees

Arctic Ocean
Atlantic Ocean
English Channel
Mediterranean Sea

Map 6: Europe (human geography)

BELGIUM	Amsterdam
DENMARK	Athens
FRANCE	Berlin
GERMANY	Bern
GREECE	Brussels
ICELAND	Copenhagen
IRISH REPUBLIC	Dublin
ITALY	Lisbon
NETHERLANDS	London
NORWAY	Madrid
POLAND	Moscow
PORTUGAL	Oslo
RUSSIA	Paris
SPAIN	Reykjavik
SWITZERLAND	Rome
UNITED KINGDOM	Warsaw

Map 7: Asia

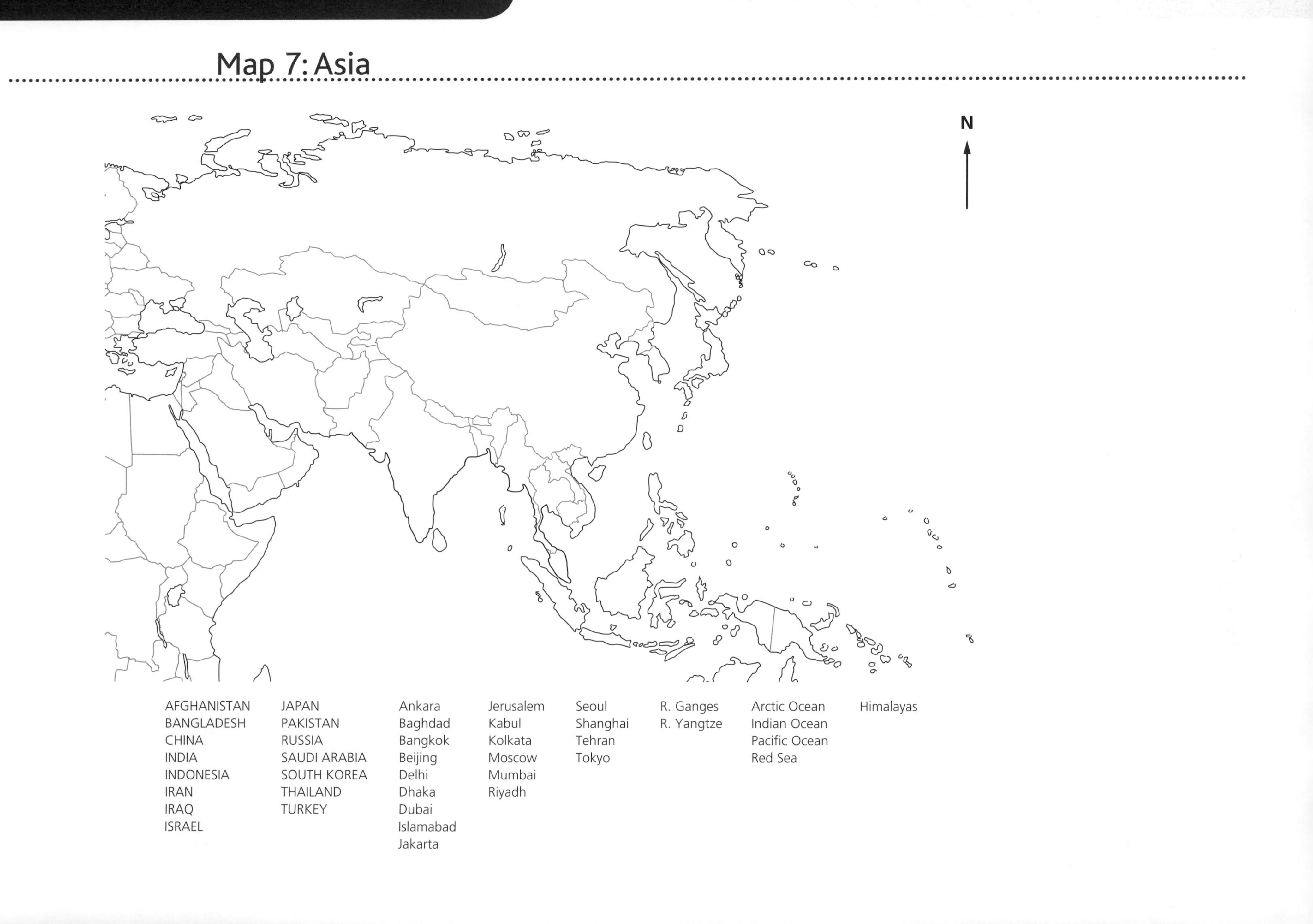

AFGHANISTAN
BANGLADESH
CHINA
INDIA
INDONESIA
IRAN
IRAQ
ISRAEL

JAPAN
PAKISTAN
RUSSIA
SAUDI ARABIA
SOUTH KOREA
THAILAND
TURKEY

Ankara
Baghdad
Bangkok
Beijing
Delhi
Dhaka
Dubai
Islamabad
Jakarta

Jerusalem
Kabul
Kolkata
Moscow
Mumbai
Riyadh

Seoul
Shanghai
Tehran
Tokyo

R. Ganges
R. Yangtze

Arctic Ocean
Indian Ocean
Pacific Ocean
Red Sea

Himalayas

Map 8: Oceania

AUSTRALIA
PAPUA NEW GUINEA
NEW ZEALAND

Canberra
Port Moresby
Sydney
Wellington

Indian Ocean
Pacific Ocean
Southern Ocean

Map 9: North and Central America

CANADA
MEXICO
USA

Los Angeles
Mexico City
New York
Ottawa
Vancouver
Washington D.C.

Rocky Mountains

Atlantic Ocean
Pacific Ocean

R. Missisipi

Map 10: South America

ARGENTINA	Bogotá	Andes	Atlantic Ocean	R. Amazon
BRAZIL	Brasilia		Pacific Ocean	
CHILE	Buenos Aires			
COLOMBIA	Lima			
PERU	Rio de Janeiro			
	Santiago			
	Sao Paulo			

Map 11: Africa

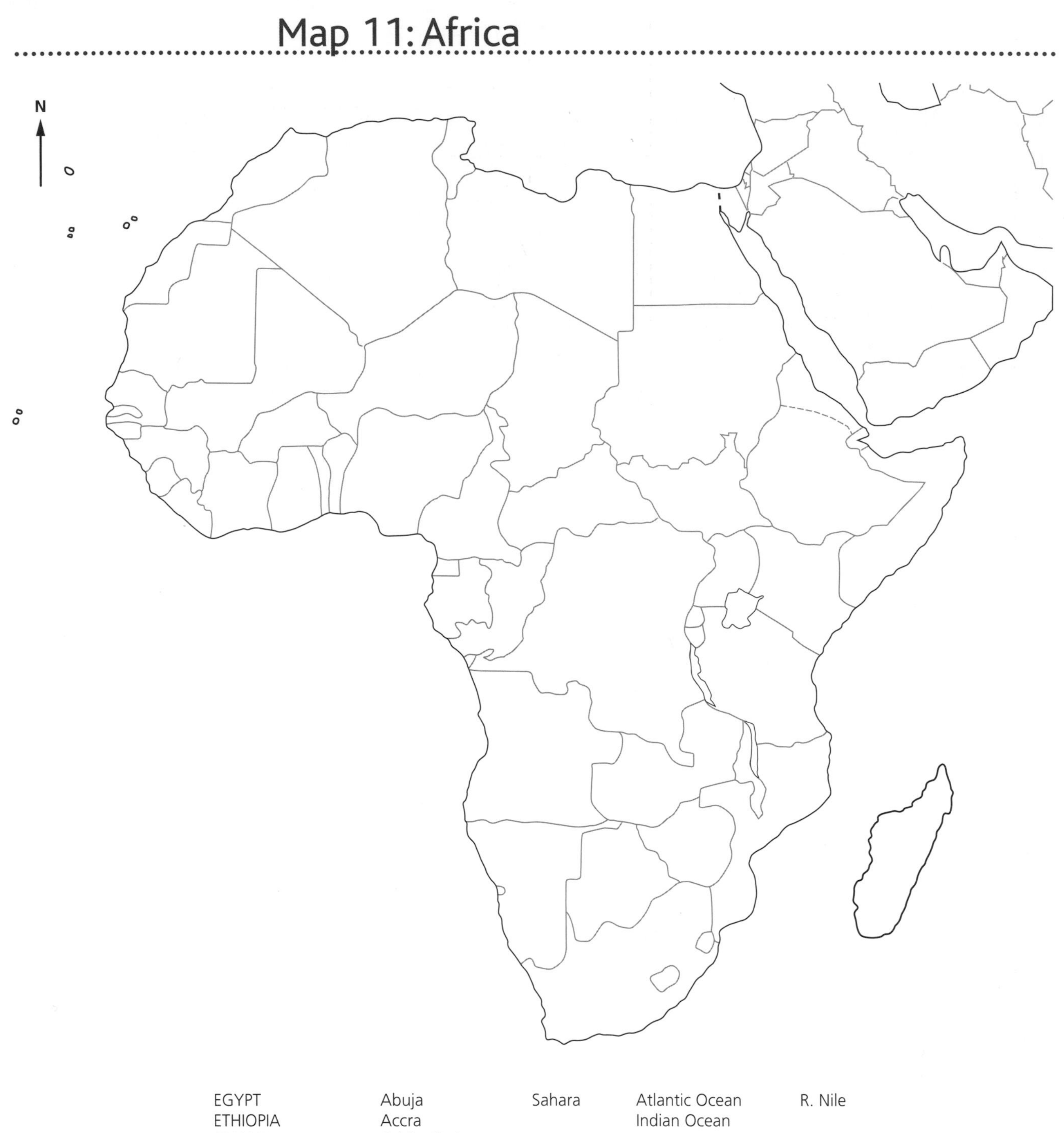

EGYPT	Abuja	Sahara	Atlantic Ocean	R. Nile
ETHIOPIA	Accra		Indian Ocean	
GHANA	Addis Ababa			
KENYA	Cairo			
NIGERIA	Nairobi			
SOUTH AFRICA	Pretoria			

Map 12: General world features

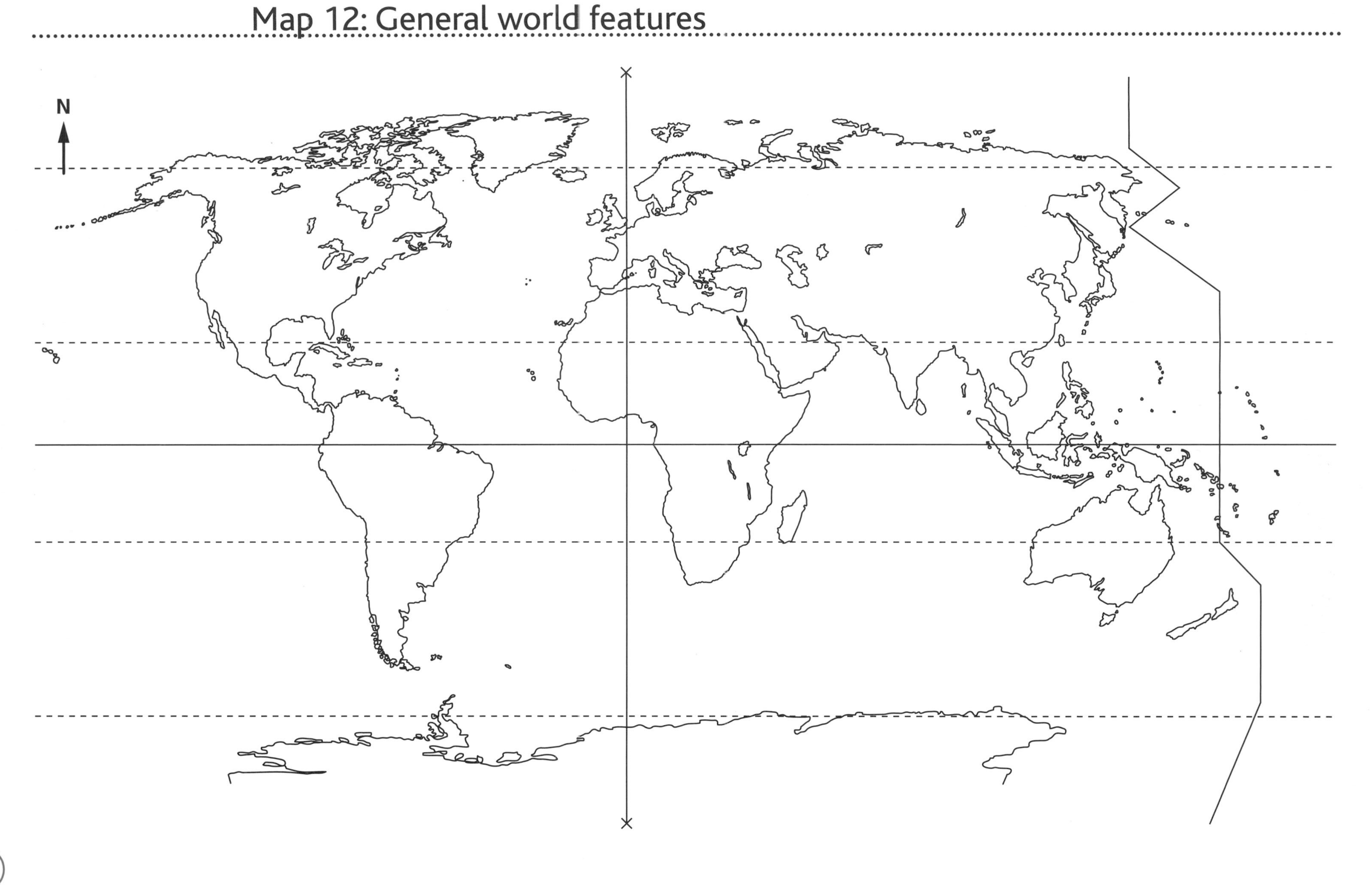